Instructor's Resource Guide

Charles Henry Brase
Regis University

Corrinne Pellillo Brase
Arapahoe Community College

Understandable Statistics

Sixth Edition

Houghton Mifflin Company   Boston   New York

Editor-in-Chief: Charles Hartford
Associate Editor: Mary Beckwith
Senior Manufacturing Coordinator: Sally Culler
Marketing Manager: Ros Kane
Editorial Assistants: Kathy Yoon and Joy Park

Printed in the U.S.A.

ISBN: 0-395-90773-X

1 2 3 4 5 6 7 8 9-SB-02 01 00 99 98

Contents

Preface

This instructor's resource guide is written to accompany the text *Understandable Statistics: Concepts and Methods*, sixth edition.

Part I contains suggestions for using the text and some alternate ways to order the topics and chapters on the basis of knowledge required for prerequisite chapters or sections are provided in the first chapter of this guide. Also included is a chart correlating the text to the *Against All Odds* video series. Finally, some suggestions for teaching Advanced Placement statistics are given.

Part II offers new chapter tests (two per chapter). The chapter tests are arranged so that they may be photocopied easily for classroom use. Each chapter test is keyed to a specific chapter, and answers with key steps are included at the end of Part II. Faculty members wishing to test over several chapters may of course choose problems from chapter tests and combine them in a suitable way.

Part III contains answers to all even-numbered problems in the text, and some of the key steps necessary to arrive at the answer. These answers also are included in the Instructor's Annotated Edition. Answers to odd-numbered problems do not appear in this guide because they are provided in Appendix III of the text. The Study and Solutions Guide for students contains complete solutions to nearly half of the odd-numbered problems.

Part IV contains transparency masters of some key diagrams from the text. In addition are some screen displays from the Texas Instruments TI-83 graphics calculator, some printouts from the computer software package *Computer Stat*, which was written to coordinate with the text *Understandable Statistics*, and some screen printouts from the professional statistical software package MINITAB.

Part V contains masters for the formula card as well as all of the tables included in Appendix II of *Understandable Statistics*, sixth edition. Using these masters, a faculty member can create custom handouts for student use during tests.

Supplements specifically designed to accompany the sixth edition of *Understandable Statistics* by Brase and Brase are

1. An *Annotated Instructor's Edition* containing margin comments and answers to even numbered problems. The answers to odd numbered problems are included in Appendix III of the Instructor's Edition as well as in the standard edition of the text used by students.

2. *Technology Guide* featuring

    Texas Instruments TI-83
    ComputerStat
    MINITAB

    This guide gives instructions for using the listed technologies as well as lab activities coordinated to the text *Understandable Statistics*.

3. ComputerStat This is a computer software package available for Windows, MS-DOS and Macintosh. This software was written by the authors Brase and Brase to coordinate with the text *Understandable Statistics*. It offers options to use built in classroom demonstrations or to enter your own data. Institutions adopting the text qualify for complimentary site licenses for this software. Contact Houghton Mifflin for details.

4. DATA DISK for MS-DOS, Windows, Macintosh   This diskette contains real world data in MINITAB portable worksheets.   The data on the disk is described in the *Using Technology Guide*. Some of the lab activities for MINITAB in the *Using Technology Guide* incorporate the data on the DATA DISK.   The same data files are also available in TI-83 graphing calculator format and can be used to link the data to the calculator.

5. *Study and Solutions Guide* by Elizabeth Farber of Bucks County Community College helps students reinforce their skills. Step by step solutions to about half of the odd-numbered problems are included. In addition Thinking About Statistics sections help students put ideas statistical ideas in context.

6. *Video Tapes* designed to accompany the text. These tapes feature Dana Mosley. They are keyed to the sections of the text.

7. Houghton Mifflin Computerized Testing for MS-DOS, Windows and Macintosh contains a computerized test bank and test design platform. Faculty can custom design their own exams utilizing test file items and their own test questions as well.

8. *Test Item File* contains a printed copy of the same questions as those included in the Computerized Testing package for *Understandable Statistics*, fifth edition.

Part I

A.   Suggestions for Using the Text
Understandable Statistics, sixth edition

B.   Alternate Paths Through the Text

C.   Correlation Chart for the *Against All Odds*
Video Series

D.   Hints for Advanced Placement Statistics Course

Suggestions for Using the Text

In writing this text we have followed the premise that a good textbook must be more than just a repository of knowledge. A good textbook should be an agent interacting with the student to create a working knowledge of the subject. To help achieve this interaction we have modified the traditional format in a way designed to encourage active student participation.

Each chapter opens with Chapter Overviews that provide non technical capsule summaries of each section to be presented in the chapter. Next is a Focus problem utilizing real world data. The Focus problem shows the student the kinds of questions they can answer when they have mastered the material in the chapter. In fact, students are asked to solve the Chapter Focus Problem as soon as concepts required for the solution have been introduced.

A special feature of this text is the occurrence of built in Guided Exercises within the reading material of the text. These Guided Exercises with their completely worked out solutions help the students focus on key concepts of newly introduced material. The Section Problems reinforce student understanding and sometimes require the student to look at the concepts from a slightly different perspective than that presented in the section. Chapter problems are much more comprehensive. They require that students place the problem in the context of all that they have learned in the chapter. Data Highlights at the end of each chapter ask students to look at data as presented in newspapers, magazines, and other media and then to apply relevant methods of interpretation. Finally, Linking Concept problems ask students to verbalize their skills and synthesize the material.

We believe that the approach from small step Guided Exercises to Section Problems to Chapter Problems to Data Highlights to Linking Concepts will enable the professor to employ his or her class time in a very profitable way, going from specific mastery details to more comprehensive decision-making analysis.

Calculators and statistical computer software relieve much of the computation burden from statistics. Many basic scientific calculators provide the mean and standard deviation. Those supporting two variable statistics provide the coefficients for the least squares line, value of the correlation coefficient, and the predicted value of y for a given x. Graphing calculators sort the data and many provide the median and quartile values. Some produce histograms, box plots, scatter plots and plot the least squares line. Statistical software packages give full support for descriptive statistics and inferential statistics. Students benefit from using these technologies. In many examples and exercises in *Understandable Statistics* we ask students to use calculators to verify answers. In text displays show TI-83 graphics calculator screens, Minitab printouts and ComputerStat printouts so that students may see the types of information readily available to them through use of technology.

However, it is not enough to enter data and punch a few buttons to get statistical results. The formulas producing the statistics contain a great deal of information about the *meaning* of the statistics. The text breaks down formulas into tabular format so that students can see the information in the formula. We find that it is useful to take class time to discuss formulas. For instance, an essential part of the standard deviation formula is the comparison of each data value to the mean. When we point this out to students, it gives meaning to the standard deviation. When students understand the content of the formulas, the numbers they get from their calculator or computer begin to make sense and have meaning.

For courses in which technologies are more incorporated into the curriculum, we provide a separate supplement, the *Technology Guide*. This guide gives specific hints for using the technologies, and specific lab activities to help students explore various statistical concepts.

New to the sixth edition is a feature called Viewpoints. These are brief essays that invite the student to consider applications of statistics in a variety of contexts. When appropriate, students are referred to Internet Web sites.

Alternate Paths Through the Text

As previous editions, the sixth edition of *Understandable Statistics* is designed to be flexible. In most one semester courses, it is not possible to present all of the topics. However, the text provides many topics so that you can tailor a course to fit your students' needs. The text provides students a *readable reference* for topics not specifically included in your course.

Table of Prerequisite Material

Chapter	Prerequisite Sections
1 Getting Started	none
2 Organizing Data	1.1
3 Averages and Variation	1.1, 2.1, 2.3
4 Elementary Probability Theory	1.1, 2.1, 2.3, 3.1, 3.2
5 The Binomial Probability Distribution and Related Topics	1.1, 2.1, 2.3, 3.1, 3.2, 4.1, 4.2 with 4.3 useful but not essential
6 Normal Distributions (omit 6.4) (include 6.4)	1.1, 2.1, 2.3, 3.1, 3.2, 4.1, 4.2, 5.1 5.2, 5.3 also
7 Introduction to Sampling Distributions	1.1, 2.1, 2.3, 3.1, 3.2, 4.1, 4.2, 5.1, 6.1, 6.2, 6.3
8 Estimation (omit 8.3 and parts of 8.4 and 8.5) (include 8.3 and all of 8.4 and 8.5)	1.1, 2.1, 2.3, 3.1, 3.2, 4.1, 4.2, 5.1, 6.1, 6.2, 6.3 7.1, 7.2 5.2, 5.3, 6.4 also
9 Hypothesis Testing (omit 9.5 and part of 9.7) (include 9.5 and all of 9.7)	1.1, 2.1, 2.3, 3.1, 3.2, 4.1, 4.2, 5.1, 6.1, 6.2, 6.3 7.1, 7.2 5.2, 5.3, 6.4 also
10 Regression and Correlation (omit part of 10.2, 10.4 and 10.5) (include all of 10.2, 10.4 and 10.5)	1.1, 2.1, 3.1, 3.2 4.1, 4.2, 5.1, 6.1, 6.2, 6.3, 7.1, 7.2, 8.1 9.1, 9.2 also

11 Chi-Square and F Distributions
    (omit 11.3)   **1.1, 2.1, 2.3, 3.1, 3.2, 4.1, 4.2, 5.1,**
                                            6.1, 6.2, 6.3, 7.1, 7.2, 9.1
    (include 11.3)   **8.1 also**

12 Nonparametric Statistics   **1.1, 2.1, 2.3, 3.1, 3.2, 4.1, 4.2, 5.1**
                                            6.1, 6.2, 6.3, 7.1, 7.2, 9.1, 9.5

Understandable Statistics
Linked to
the Annenberg/CBS Video Series
Against All Odds: Inside Statistics

The television series *Against All Odds: Inside Statistics* produced by Annenberg/CBS consists of 26 programs, each 30 minutes in length. The video programs provide interesting applications of statistics on actual location. Some of the programs expand the discussion of topics found in the text *Understandable Statistics*.

Against All Odds videos correlate to the text *Understanding Basic Statistics* in the following way.

Understanding Basic Statistics Chapter	*Against All Odds* Program
Chapter 1 Getting Started	Program 1 What is Statistics?

Chapter 2 Organizing Data
  2.1 Random Samples
 	
  2.2 Graphs
  2.3 Histograms and Frequency Distributions
  2.4 Stem-and-Leaf Displays
	Program 12 Experimental Design
	Program 13 Experiments and Samples
	
	Program 2 Picturing Distributions

Chapter 3 Averages and Variation
  3.1 Measures of Central Tendency: Mode, Median, and Mean
  3.2 Measures of Variation
  3.3 Box-and-Whisker Plot
	Program 3 Numerical Description of Distributions

Chapter 4 Introduction to Probability
  4.1 What is Probability?
  4.2 Some Probability Rules
  4.3 Trees and Counting Techniques
	Program 15 What is Probability?

Chapter 5   The Binomial Probability Distribution and Related Topics
  5.1 Introduction to Random Variables and Probability Distributions
 	
  5.2 Binomial Probabilities
  5.3 The Mean and Standard Deviation of the Binomial Distribution
	Program 16 Random Variables
	
	Program 17 Binomial Distributions (first part)

Chapter 6 Normal Distributions 6.1 Graphs of Normal Probability Distributions	Program 4 Normal Distributions
6.2 Standard Units and the Standard Normal Distribution	Program 5 Normal Calculations
6.3 Areas Under Any Normal Curves	Note: Table referred to gives areas in the tails of the distribution rather than from 0 to z
6.4 Normal Approximation to the Binomial Distribution	Program 17 Binomial Distributions (last part)
Chapter 7 Introduction to Sampling Distributions 7.1 Sampling Distributions	Program 14 Samples and Sampling Distributions
7.2 The Central Limit Theorem	Program 18 The Sample Mean and Control Charts Note: These are $\bar{x}$ control charts rather than x control charts as discussed in Section 6.1 of the text.

Chapter 8 Introduction to Estimation
   8.1 Estimating μ with Large Samples
   8.4 Choosing the Sample Size
	Program 19 Confidence Intervals

Chapter 9 Hypothesis Testing Involving One Population 9.1 Introduction to Hypothesis Testing	Program 20 Significance Tests
9.2 Tests Involving the Mean μ (Large Samples) 9.3 The P value in Hypothesis Testing	Program 21 Inference for One Mean Note: Whenever σ is unknown the t distribution is used even with large samples
9.4 Tests Involving the Mean μ (Small Samples)	
9.5 Tests Involving a Proportion	Program 23 Inference for Proportions (first part)
9.6 Testing Differences of Two Means or two Proportions	Program 22 Comparing Two Means Note: Program uses the t distribution if either σ_1 or σ_2 is not known Program 23 Inference for Proportions (last part)

Chapter 10 Regression and Correlation
    10.1   Introduction to Paired Data and
                 Scatter Diagrams
    10.2   Linear Regression and Confidence
                 Bounds for Prediction
	Program 7 Models for Growth
	Program 8 Describing Relationships

10.3 The Linear Correlation Coefficient	Program 9 Correlation Program 11 The Question of Causation
10.4 Testing the Correlation Coefficient	Program 25 Inference for Relationships
10.5 Multiple Regression	Program 10 Multidimensional Data Analysis

Chapter 11 Chi-Square and F Distributions
    11.1 Chi-Square : Tests of Independence
	Program 24 Inference for Two-Way
	                      Tables

Summary of Course	Program 26 Case Study

Hints for Advanced Placement Statistics Course

Text:

Understandable Statistics provides a strong text for the main concepts included on the AP Examination in Statistics.

Calculator:

Each students is expected to use a graphing calculator with statistics capabilities during the AP exam. A calculator such as the Texas Instruments TI-83 is particularly useful.

Outline of Topics:

The major topics covered by the AP Examination in Statistics are as follows.

I. Exploring Data: Observing patterns and departure from patterns
   Use of graphical and numerical techniques to study patterns and departure from patterns with an emphasis on interpreting the information from graphical displays and numerical summaries.
   A. Single variable data
      Students should be able to use graphs such as histograms, stem-and-leaf plots, box-and-whisker plots to get information about the shape of the data distribution, variability of the data, and unusual values. Summary statistics such as the mean, median, mode give measures of center. Values such as the variance, standard deviation, range, interquartile range describe the spread of the data. Quartiles, percentiles, standardized scores (z-scores) give information about position. Students should be able to compare data distributions based upon graphs and summary statistics. In addition, they should know the effect of changing units on summary measures.
   B. Data Pairs
      Students should be able to use scatter plots to detect patters for data pairs and find the equation of the least squares regression line. The correlation coefficient and residual plots can be used to determine the strength of the linear model. Students should be able to identify outliers and influential points. Relation of causation to correlation should be understood.
   C. Categorical data tables

II. Planning a Study: deciding what and how to measure
   In open ended questions students need to demonstrate that they understand how to gather data according to a well-developed plan. Methods of data collection such as census, sample survey using simple random samples, experiment, observational study may be used. Students should demonstrate an awareness of sampling error, sources of bias, and use of stratifying to reduce variation. Experimental design such as completely randomized design for two treatments and blocking designs might be required.

III. Anticipating Patterns: Producing models using probability and simulation
In this topic students use probability as a tool for anticipating what the distribution of data should look like under a given model. Particular topics include rules of probability, concept of law of large numbers, independence of events, mean and standard deviation of a random variable, use of binomial distribution, normal distribution and sampling distributions of a sample proportion, sample mean, difference of two independent sample proportions, difference between two independent sample means.

IV. Statistical Inference: Confirming models
Select appropriate models for statistical inference
A. Confidence intervals: meaning of; large sample confidence interval for a proportion; large sample confidence interval for a mean; large sample confidence interval for a difference between two proportions; large sample confidence interval for a difference between two means (unpaired and paired)
B. Tests of significance: structure and logic of test including null hypothesis, alternate hypothesis, p values, one-and two-tailed tests; large sample test for a proportion, large sample test for a mean; large sample test for a proportion; large sample test for a mean; large sample test for a difference between two proportions; large sample test for a difference between two means (unpaired and paired); Chi-square test for goodness of fit, homogeneity of proportions, and independence
C. Special case of normally distributed data: t-distribution; single sample t procedures; two sample (independent and matched pairs) t procedures; inference for slope of least squares line.

Type of Questions

The test has two parts. Multiple-choice questions comprise one part. The second part is a free-response section requiring the student to answer open-ended questions and to complete an investigative task involving more extended reasoning. The two sections are given equal weight in determining the grade for the examination.

Duration of the Test

The test is three hours long.

Formulas and Tables

Formulas and tables are provided for students taking the AP Statistics Examination. The format of the formulas and some of the tables is slightly different from those in the text. Note that the normal distribution table for the AP exam gives areas in the left tail of the distribution. The Student's t distribution table provides critical values for areas in the right tail of the distribution, and gives critical values for different confidence levels. Copies of the formulas and tables available during the AP exam are included in the Advanced Placement Course Description for Statistics.

More Information and Sample Questions

Advanced Placement Course Description Statistics is a publication by the College Board that contains sample questions, formulas, tables, and a description of the AP Statistics course. The publication is available from the College Board

> The College Board
> 45 Columbus Avenue
> New York, NY 10023-6992
>
> (phone: 212-713-8000)

Part II

Chapter Tests

Answers to Chapter Tests

Part II contains sample tests and answers for each chapter. Additional test items may be found in a separate manual, Test Item File, that is available in print copy and also as computerized testing software in MS-DOS, Windows, and Macintosh platforms.

Chapter Test 1A

1. The Colorado State Legislature wants to estimate the length of time (semesters) it takes a resident of Colorado to earn a Bachelor's degrees from a state college or university. A random sample of 265 recent (in state) graduates were surveyed.
    (a) What is the implied population?
    (b) What is the sample?
    (c) Would this sample necessarily reflect the length of time (semesters) for foreign students to earn a degree at a Colorado university or college. Explain your answer.

2. A student advising file contains the following information.
    (a) Name of student
    (b) Student I.D. number
    (c) Cumulative grade point average
    (d) Dates of awards (scholarships, dean's list,...)
    (e) Declared major or undecided if no major declared
    (f) A number code representing class standing: 1 = Freshman, 2 = Sophomore, 3 = Junior, 4 = Senior, 5 = Graduate student
    (g) Entrance exam rating for competency in English: Excellent, Satisfactory, Unsatisfactory
    For the information in parts a through g, list the highest level of measurement as ratio, interval, ordinal, or nominal and explain your choice.

3. Categorize the style of gathering data (sampling, experiment, simulation, census) described in each of the following situations:
    (a) Look at all the apartments in a complex and determine the monthly rent charged for each unit.
    (b) Given one group of students a flu vaccination and compare the number of times these students are sick during the semester with students in a group who did not receive the vaccination.
    (c) Select a sample of students and determine the percentage who are taking mathematics this semester.
    (d) Use a computer program to show the effects on traffic flow when the timing of stop lights is changed.

4. Write a brief essay in which you describe what is meant by an experiment. Give an example of a situation in which data is gathered by means of an experiment. How is gathering data from an experiment different from using a sample from a specified population?

Chapter Test 1B

1. A book store wants to estimate the proportion of its customers who buy murder mysteries. A random sample of 76 customers are observed at the checkout counter and the number purchasing murder mysteries is recorded.
    (a) What is the implied population?
    (b) What is the sample?
    (c) Would this sample necessarily reflect the proportion of customers purchasing non-fiction books? Explain your answer.

2. A restaurant manager is developing a clientele profile. Some of the information for the profile follows:
    (a) Gender of diners.
    (b) Size of groups dining together
    (c) Time of day the last diner of the evening departs
    (d) Age grouping: young, middle age, senior
    (e) Length of time a diner waits for a table
    For the information in parts (a) though (e), list the highest level of measurement as ratio, interval, ordinal, or nominal and explain your choice.

3. Categorize the style of gathering data (sampling, experiment, simulation, census) for the following situations:
    (a) Consider all the students enrolled at your college this semester and report the age of each student.
    (b) Select a sample of new F10 pickup trucks and count the number of manufacturer defects in each of the trucks.
    (c) Use computer graphics to determine the flight path of a golf ball when the position of the hand on the golf club is changed.
    (d) Teach one section of English composition using a specific word processing package and teach another without using any computerized word processing. Count the number of grammar errors made by students in each section on a final draft of a 20 page term paper.

4. Write a brief essay in which you discuss some of the aspects of surveys. Give specific examples to illustrate your main points.

Chapter Test 2A

1. Consider the experiment of rolling a single die. Describe how you would use a random number table to simulate the outcomes of rolling a single die. Using the following row of random numbers from the table, find the first five outcomes.

   36017 98590 64180 72315 39710

2. The Dean's Office at Hendrix College gave the following information about numbers of majors in different academic areas: Humanities, 372; Natural Science, 415; Social Science, 511; Business Administration, 619; Philosophy, 196. Make a Pareto chart representing this information.

3. Professor Hill in the Music Department kept a list of the number of students visiting his office each week for two semesters (30 weeks). The results were

15	23	17	13	3	9	7	6	8	11
16	32	27	4	20	3	28	5	6	11
20	12	8	10	25	10	8	15	11	9

   (a) Make a frequency table with five classes, showing class boundaries, class midpoints, frequencies, relative frequencies, and cumulative frequencies.
   (b) Make a frequency histogram with five classes.
   (c) Make a relative frequency histogram with five classes.
   (d) Make an ogive with five classes.

4. Jim is a taxi driver who keeps a record of his meter readings. The results for the past twenty meter readings (rounded to the nearest dollar) are given below.

15	7	9	21	19	17	8	35	22	33
46	5	24	37	51	49	57	42	12	16

   Make a stem-and-leaf display of the data.

5. Identify each of the following samples by naming the sampling technique used (cluster, convenience, random, stratified, systematic).
   (a) Measure the length of time every fifth person coming into a bank waits for teller service over a period of two days.
   (b) Take a sample of five zip codes from the Chicago metropolitan region and use all the elementary schools from each of the zip code regions. Determine the number of students enrolled in first grade in each of the schools selected.
   (c) Divide the users of the computer online service Internet into different age groups and then select a random sample from each age group to survey about the amount of time they are connected to Internet each month.
   (d) Survey five friends regarding their opinion of the student cafeteria.
   (e) Pick a random sample of students enrolled at your college and determine the number of credit hours they have each accumulated toward their degree program.

Chapter Test 2A continued

6. The Air Pollution Index in Denver for each day of the second week of February is shown below.

    1.7   2.4   5.3   4.1   3.2   2.0   2.5

    Make a time plot for these data.

Chapter Test 2B

1. A business employs 736 people. Describe how you could get a simple random sample of size 30 to survey regarding desire for professional training opportunities. Identify the first 5 to be included in the sample using the following random number sequence

   62283   14130   55790   40133   47596   17654

2. A book store recorded the following sales last month by genre: Romance, 519; Murder Mystery, 732; Biography, 211; Self help, 819; Travel guide, 143; Children's books, 643. Make a Pareto chart displaying this information.

3. The College Registrar's Office recorded the number of students receiving a grade of Incomplete. Results for the past 24 quarters are

28	47	19	58	63	77	53	39	93	35
42	81	62	67	71	59	48	56	75	48
63	32	46	57						

   (a) Make a frequency table with five classes, showing class boundaries, class midpoints, frequencies, relative frequencies, and cumulative frequencies.
   (b) Make a frequency histogram with five classes.
   (c) Make a relative frequency histogram with five classes.
   (d) Make an ogive with five classes.

4. The Humanities Division recorded the number of students signed up for the Study Abroad Program each quarter. The results are

58	26	21	29	33	47	42	38	44	56
52	64	68	59	63	36	34	45	51	50

   Make a stem-and-leaf display of the data.

5. To determine monthly rental prices of apartment units in the San Francisco area, samples were constructed in the following ways. Categorize (cluster, convenience, simple random, stratified, systematic) each sampling technique described.
   (a) Number all the units in the area and use a random number table to select the apartments to include in the sample.
   (b) Divide the apartment units according to number of bedrooms and then sample from each of the groups.
   (c) Select 5 zip codes at random and include every apartment unit in the selected zip codes.
   (d) Look in the newspaper and consider the first sample of apartment units that list rent per month.
   (e) Call every 50th apartment complex listed in the yellow pages and record the rent of the unit with unit number closest to 200.

Chapter Test 2B continued

6. The price of gold (1 troy ounce) was recorded each week for twelve weeks. The data are rounded to the nearest dollar.

289	291	298	305	311	322
316	300	290	299	291	288

Make a time plot for these data.

Chapter Test 3A

1. A random sample of 18 airline carry-on luggage bags gave the following weights (rounded to the nearest pound).

12	25	10	38	12	19	8	12	17
41	7	22	10	19	12	16	5	14

   Find the mean, median, and mode of these weights.

2. A random sample of 7 Northern Pike from Taltson Lake (Canada) gave the following lengths rounded to the nearest inch.

21	27	46	35	41	36	25

   (a) Find the range.
   (b) Find the sample mean.
   (c) Find the sample variance.
   (d) Find the standard deviation.

3. A random sample of receipts for individuals eating at the Terrace Restaurant showed the sample mean to be $\bar{x} = \$10.38$ with sample standard deviation $s = \$2.17$.

   (a) Compute the coefficient of variation for this data.
   (b) Use Chebyshev's Theorem to find the smallest interval centered on the mean in which we can expect at least 75% of the data to fall.

4. A random sample of 330 adults were asked the maximal amount (dollars) they would spend on a ticket to a top rated performance. The results follow where x is the cost and f is the number of people who would spend that maximal amount.

x	20	30	40	50	60
f	62	83	120	40	25

   (a) Compute the sample mean.
   (b) Compute the sample variance.
   (c) Compute the sample standard deviation.

5. A random sample of 27 skiers at Vail, Colorado gave their ages. The results were

18	25	32	16	41	53	29	58	23
62	47	56	19	22	38	15	46	33
49	52	37	26	72	44	19	24	29

   (a) Give the five number summary including the low value, Q_1, median, Q_3, and high value.
   (b) Make a box-and-whisker plot for the given data.
   (c) Find the interquartile range.

Chapter Test 3A continued

6. In Biology 340, weights are assigned to required activities as follows:

    project, 25%; exam 1, 15%; exam 2, 15%; exam 3, 15%; final exam, 30%.

    Each activity is graded a 100 point scale. Gary earned 75 points on the project, 85 points on exam 1, 95 points on exam 2, 90 points on exam 3, and 88 points on the final exam. Compute his overall weighted average in the Biology 340 class.

Chapter Test 3B

1. A veterinarian in a small animal clinic had the following record of life spans of Golden Retrievers (to the nearest year).

9	12	15	11	8	10	7	5	11	14
13	6	11	16	11	14	11	4	12	11

    Find the mean, median, and mode for this data.

2. A random sample of 6 people, each 20 pounds overweight, volunteered to go on the same diet. After 3 months, their weight loss (pounds) were

12	5	14	19	15	8

    (a) Find the range.
    (b) Find the sample mean.
    (c) Find the sample variance.
    (d) Find the sample standard deviation.

3. A large sample of Northern Pike caught at Taltson Lake (Canada) showed that the average length was $\bar{x} = 32.5$ inches with sample standard deviation $s = 8.6$ inches.
    (a) Compute the coefficient of variation for this data.
    (b) Use Chebyshev's Theorem to find an interval centered on the mean in which we can expect at least 75% of the data to fall.

4. A random sample of 146 students in Chemistry 215 gave the following grade information (A = 4.0, B = 3.0, C = 2.0, D = 1.0, and F = 0). In the following table x = grade and f = number of students receiving this grade.

x	0	1	2	3	4
f	8	14	62	43	19

    (a) Find the sample mean.
    (b) Find the sample variance.
    (c) Find the sample standard deviation.

5. A random sample of 24 professors at Montana State University gave the following ages (years)

29	32	56	61	27	43	38	65
36	47	41	68	59	40	33	35
44	39	28	46	42	62	58	45

    (a) Give the five number summary including the low value, Q_1, median, Q_3, and high value.
    (b) Make a box-and-whisker plot for the data.
    (c) Find the interquartile range.

Chapter Test 3B continued

6. A teacher evaluation rating system uses the following items and weights:

   Availability of professor outside of class, 10%
   Clarity of presentation, 25%
   Respectful of student ideas, 20%
   Grades fairly, 15%
   Knowledge of the field,   30%

Each item is rated on a 100 point scale. Dr. Gill was evaluated by her Sociology 350 class and received the following ratings:

 Availability, 65; Clarity, 85; Respects students, 90; Grades fairly, 70; Knowledge, 95

Compute a weighted average to determine Dr. Gill's overall teacher rating.

Chapter Test 4A

1. A random sample of 317 new Smile Bright electric toothbrushes showed 19 were defective.
   (a) How would you estimate the probability that a new Smile Bright electric toothbrush is defective? What is your estimate?
   (b) What is your estimate for the probability that a Smile Bright electric toothbrush is not defective?
   (c) Either an electric toothbrush is defective or not. What is the sample space in this problem? Do the probabilities assigned to the sample space add up to one?

2. An urn contains 12 balls identical in every respect except color. There are 3 red balls, 7 green balls, and 2 blue balls.
   (a) You draw two balls from the urn, but replace the first ball before drawing the second. Find the probability that the first ball is red and the second is green.
   (b) Repeat part a, but do not replace the first ball before drawing the second.

3. Robert is applying for a bank loan to open up a pizza franchise. He must complete a written application, and then be interviewed by bank officers. Past records for this bank show that the probability of being approved in the written part is 0.63. Then the probability of being approved by the interview committee is 0.85, given the candidate has been approved on the written application. What is the probability Robert is approved on both the written application and the interview?

4. A hair salon did a survey of 360 customers regarding satisfaction with service and type of customer. A walk-in customer is one who has seen no ads and not been referred. The other customers either saw a TV ad or were referred to the salon (but not both). The results follow.

	Walk In	TV Ad	Referred	Total
Not Satisfied	21	9	5	35
Neutral	18	25	37	80
Satisfied	36	43	59	138
Very Satisfied	28	31	48	107
Total	103	108	149	360

   Assume the sample represents the entire population of customers. Find the probability that a customer is
   (a) Not satisfied
   (b) Not satisfied <u>and</u> walk in
   (c) Not satisfied, <u>given</u> referred
   (d) Very satisfied
   (e) Very satisfied, <u>given</u> referred
   (f) Very satisfied <u>and</u> TV ad
   (g) Are the events satisfied and referred independent or not? Explain your answer.

5. A computer package sale comes with two different choices of printers and four different choices of monitors. If a store wants to display each package combination that is for sale, how many packages must be displayed? Make a tree diagram showing the outcomes for selecting printer and monitor.

Chapter Test 4B

1. The Student Council is made up of 4 women and 6 men. One of the women is president of the Council. A member of the council is selected at random to report to the Dean of Student Life.
   (a) What is the probability a woman is selected?
   (b) What is the probability a man is selected?
   (c) What is the probability the president of the Student Council is selected?

2. An urn contains 17 balls, identical in every respect except color. There are 6 red balls, 8 green balls, and 3 blue balls.
   (a) You draw two balls from the urn, but replace the first before drawing the second. Find the probability that the first ball is red and the second ball is green.
   (b) Repeat part a, but do not replace the first ball before drawing the second.

3. The Dean of Hinsdale College found that 12% of the female students are majoring in Computer Science. If 64% of the students at Hinsdale are women, what is the probability that a student chosen at random will be a woman majoring in Computer Science?

4. The Committee on Student Life did a survey of 417 students regarding satisfaction with Student Government and class standing. The results follow:

	Freshman	Sophomore	Junior	Senior	Total
Not Satisfied	17	19	23	12	71
Neutral	61	35	32	38	166
Satisfied	23	49	43	65	180
Total	101	103	98	115	417

Assume the sample represents the entire po8ulation of students. Find the probability that a student selected at random is
   (a) Satisfied (with Student Government)   (b) Satisfied, given the student is a Senior
   (c) Neutral   (d) Neutral and freshman
   (e) Neutral, given the student is a Freshman
   (f) Senior, given satisfied
   (g) Are the events, Freshman and neutral independent or not? Explain.

5. A fishing camp has 16 clients. Each cabin at the camp will accommodate 5 fishermen. In how many different ways can the first cabin be filled with clients?

Chapter Test 5A

1. Sam is a representative who sells large appliances such as refrigerators, stoves, and so forth. Let x = number of appliances Sam sells on a given day. Let f = frequency (number of days) with which he sells x appliances. For a random sample of 240 days, Sam had the following sales record.

x	0	1	2	3	4	5	6	7
f	9	72	63	41	28	14	8	5

Assume the sales record is representative of the population of all sales days
   (a) Use relative frequency to find P(x) for x = 0 to 7.
   (b) Use a histogram to graph the probability distribution of part a.
   (c) Compute the probability that x is between 2 and 5 (including 2 and 5).
   (d) Compute the probability that x is less than 3.
   (e) Compute the expected value of the x distribution.
   (f) Compute the standard deviation of the x distribution.

2. Of those mountain climbers who attempt Mt. McKinley (Denali), only 65% reach the summit. In a random sample of 16 mountain climbers who are going to attempt Mt. McKinley, what is the probability that
   (a) All 16 reach the summit?
   (b) At least 10 reach the summit?
   (c) No more than 12 reach the summit?
   (d) From 9 to 12 reach the summit, including 9 and 12.

3. The probability that a truck will be going over the speed limit on I-80 between Cheyenne and Rock Springs, Wyoming is about 75%. Suppose a random sample of 5 trucks on this stretch of I-80 are observed.
   (a) Make a histogram showing the probability that r = 0, 1, 2, 3, 4, 5 trucks going over the speed limit.
   (b) Find the mean μ of this probability distribution.
   (c) Find the standard deviation of the probability distribution.

4. Records show that the probability of catching a Northern Pike over 40 inches at Taltson Lake (Canada) is about 15% for each full day a person spends fishing. What is the minimal number of days a person must fish to be at least 83.3% sure of catching one or more Northern Pike over 40 inches.

5. The probability that an airplane is more than 45 minutes late on arrival is about 15%. Let n = 1, 2, 3, ... represent the number of times a person travels on an airplane until the first time the plane is more than 45 minutes late.
   (a) Write a brief but complete discussion in which you explain why the Geometric distribution would be appropriate. Write out a formula for the probability of the random variable n.
   (b) What is the probability that the 3rd time a person flies, he or she is on a flight that is more than 45 minutes late?
   (c) What is the probability that more than three flights are required before a plane is more than 45 minutes late?

Chapter Test 5A continued

6. Suppose the average number of customers entering a store in a 20 minute period is 6 customers. The store wants a probability distribution for the number of people entering the store each 20 minutes.
   (a) Write a brief but complete discussion in which you explain why the Poisson distribution would be appropriate. What is λ? Write out the formula for the probability distribution.
   (b) What is the probability that exactly 3 customers enter the store during a 20 minute period?
   (c) What is the probability that more than 3 customers enter the store during a 20 minute period?

7. The probability a new medication will cause a bad side effect is 0.03. The new medication has been given to 150 volunteers. Let r be the random variable representing the number of people who have a bad side effect.
   (a) Write a brief but complete discussion in which you explain why the Poisson approximation to the binomial would be appropriate. Are the assumptions satisfied? What is λ? Write out a formula for the probability distribution of r.
   (b) Compute the probability that exactly 3 people from the sample of 150 volunteers will have a bad side effect from the medication?
   (c) Compute the probability that more than 3 people out of the sample of 150 volunteers will have a bad side effect from the medication?

Chapter Test 5B

1. An aptitude test was given to a random sample of 228 people intending to become Data Entry Clerks. The results are shown below where x is the score on a 10 point scale, and f is the frequency of people with this score.

x	1	2	3	4	5	6	7	8	9	10
f	15	21	46	51	42	18	12	10	8	5

Assume the above data represents the entire population of people intending to become Data Entry Clerks.
(a) Use relative frequencies to find P(x) for x = 1 to 10.
(b) Use a histogram to graph the probability distribution of part a.
(c) To be accepted into a training program, students must have a score of 4 or higher. What is the probability an applicant selected at random will have this score?
(d) To receive a tuition scholarship a student needs a score of 8 or higher. What is the probability an applicant selected at random will have such a score?
(e) Compute the expected value of the x distribution.
(f) Compute the standard deviation of the x distribution.

2. Of all college freshmen who try out for the track team, the coach will only accept 30%. If 15 freshmen try out for the track team, what is the probability that
(a) all 15 are accepted?
(b) at least 8 are accepted?
(c) no more than 4 are accepted?
(d) between 5 and 10 are accepted (including 5 and 10).

3. The probability a vehicle will change lanes while making a turn is 55%. Suppose a random sample of 7 vehicles are observed making turns at a busy intersection.
(a) Make a histogram showing the probability that r = 0, 1, 2, 3, 4, 5, 6, 7 vehicles will make a lane change while turning.
(b) Find the expected value μ of this probability distribution.
(c) Find the standard deviation of this probability distribution.

4. Past records show that the probability of catching a Lake Trout over 15 pounds at Talston Lake (Canada) is about 20% for each full day a person spends fishing. What is the minimal number of days a person must fish to be at least 89.3% sure of catching one or more Lake Trout over 15 pounds?

5. Past records at an appliance store show that about 60% of the customers who look at appliances will buy one. Let n = 1, 2, 3, ... represent the number of customers a sales clerk must help until the <u>first</u> sale of the day.
(a) Write a brief but complete discussion in which you explain why the Geometric Distribution would apply in this context. Write out a formula for the random variable n.
(b) Compute P(n = 4).
(c) Compute P(n $\geq$ 3).

Chapter 5B Test Continued

6. At Community Hospital maternity ward, babies arrive at an average of 8 babies per hour. The hospital staff wants a probability distribution for the number of babies arriving each hour.
    (a) Write a brief but complete discussion in which you explain why the Poisson distribution would be appropriate. What is λ? Write out the formula for the probability distribution.
    (b) What is the probability exactly 7 babies are born during the next hour?
    (c) What is the probability fewer than 3 babies are born during the next hour?

7. As a telecommunications satellite goes over the horizon, stored messages are relayed to the next satellite which is still in position. However, the probability is 0.01 that an interruption will occur, and the relay transmission will be lost. Out of 200 such relays, let r be the random variable that represents the number of transmissions that are lost.
    (a) Write a brief but complete discussion in which you explain why the Poisson approximation to the binomial would be appropriate. Are the assumptions satisfied? What is λ? Write out a formula for the probability distribution of r.
    (b) Compute the probability that exactly two transmissions are lost.
    (c) Compute the probability that more than two transmissions are lost.

Chapter Test 6A

1. Let x be a random variable that represents the length of time it takes a student to complete Dr. Gill's Chemistry Lab Project. From long experience, it is known that x has a normal distribution with mean $\mu = 3.6$ hours and standard deviation $\sigma = 0.5$ hours.
    Convert each of the following x intervals to standard z intervals.
    (a) $x \geq 4.5$ (b) $3 \leq x \leq 4$ (c) $x \leq 2.5$
    Convert each of the following z intervals to raw score x intervals.
    (d) $z \leq -1$ (e) $1 \leq z \leq 2$ (f) $z \geq 1.5$

2. John and Joel are salesmen in different districts. In John's district, the long term mean sales is $17,319 each month with standard deviation $684. In Joel's district, the long term mean sales is $21,971 each month with standard deviation $495. Assume that sales in both district follow a normal distribution.
    (a) Last month John sold $19,214 whereas Joel sold $22,718 worth of merchandise. Relative to the buying habits of customers in each district, does this mean Joel is a better salesman? Explain.
    (b) Convert Joel's sales last month to a standard z score, and do the same for John's sales last month. Then locate both z scores under a standard normal curve. Who do you think is the better salesman? Explain your answer.

3. The length of time to complete a door assembly on an automobile factory assembly line is normally distributed with mean $\mu = 6.7$ minutes and standard deviation $\sigma = 2.2$ minutes. For a door selected at random, what is the probability the assembly time will be
    (a) 5 minutes or less?
    (b) 10 minutes or more?
    (c) between 5 and 10 minutes?

4. From long experience, it is known that the time it takes to do an oil change and lubrication job on a vehicle has a normal distribution with mean $\mu = 17.8$ minutes and standard deviation $\sigma = 5.2$ minutes. An auto service shop will give a free lube job to any customer who must wait beyond the guaranteed time to complete the work.. If the shop does not want to give more than 1% of its customers a free lube job, how long should the guarantee be (round to the nearest minute).

Chapter Test 6A continued

5. You are examining a quality control chart regarding number of employees absent each shift from a large manufacturing plant. The plant is staffed so that operations are still efficient when the average number of employees absent each shift is $\mu = 15.7$ with standard deviation $\sigma = 3.5$. For the most recent 12 shifts, the number of absent employees were

Shift	1	2	3	4	5	6	7	8	9	10	11	12
#	6	10	7	16	19	18	17	21	22	18	16	19

   (a) Make a control chart showing the number of employees absent during the 12 day period.
   (b) Are there any periods during which the number absent is out of control? Identify the out of control periods according to Type I, Type II, Type III out of control signals.

6. Medical treatment will cure about 87% of all people who suffer from a certain eye disorder. Suppose a large medical clinic treats 57 people with this disorder. Let r be a random variable that represents the number of people that will recover. The clinic wants a probability distribution for r.
   (a) Write a brief but complete description in which you explain why the normal approximation to the binomial would apply. Are the assumptions satisfied? Explain.
   (b) Estimate $P(r \leq 47)$.
   (c) Estimate $P(47 \leq r \leq 55)$.

Chapter Test 6B

1. Let x be a random variable that represents the length of time it takes a student to write a term paper for Dr. Adam's Sociology class. After interviewing many students, it was found that x has an approximately normal distribution with mean $\mu = 6.8$ hours and standard deviation $\sigma = 2.1$ hours.
   Convert each of the following x intervals to standardized z units.
      (a) $x \leq 7.5$ (b) $5 \leq x \leq 8$ (c) $x \geq 4$
   Convert each of the following z intervals to raw score x intervals.
      (d) $z \geq -2$ (e) $0 \leq z \leq 2$ (f) $z \leq 3$

2. Operating temperatures of two models of portable electric generators follow a normal distribution. For generator I, the mean temperature is $\mu_1 = 148°F$ with standard deviation $\sigma_1 = 25°F$. For generator II, the mean temperature is $\mu_2 = 143°F$ with standard deviation $\sigma_2 = 8°F$. At peak power demand, generator I was operating at $166°F$, and generator II was operating at $165°F$.
   (a) At peak power output both generators are operating at about the same temperature. Relative to the operating characteristics, is one a lot hotter than the other? Explain.
   (b) Convert the peak power temperature for each generator to standard z units. Then locate both z scores under a standard normal curve. Could one generator be near a melt down? Which one? Explain your answer.

3. Weights of a certain model of fully loaded gravel trucks follow a normal distribution with mean $\mu = 6.4$ tons and standard deviation $\sigma = 0.3$ tons. What is the probability that a fully loaded truck of this model is
   (a) less than 6 tons?
   (b) more than 7 tons?
   (c) between 6 and 7 tons?

4. Quality control studies for Speedy Jet Computer Printers show the lifetime of the printer follows a normal distribution with mean $\mu = 4$ years and standard deviation $\sigma = 0.78$ years. The company will replace any printer that fails during the guarantee period. How long should Speedy Jet printers be guaranteed if the company wishes to replace no more than 10% of the printers?

5. A toll free computer software support service for a spreadsheet program has established target length of time for each customer help phone call. The calls are targeted to have mean duration of 12 minutes with standard deviation 3 minutes. For one help technician the most recent 10 calls had the following duration.

call #	1	2	3	4	5	6	7	8	9	10
Length	15	25	10	9	20	19	11	5	4	8

   (a) Make a control chart showing the lengths of calls
   (b) Are there any periods during which the length of calls are out of control? Identify the out of control periods and comment about possible causes.

Chapter Test 6B continued

6. Psychology 231 can be taken as a correspondence course on a Pass/Fail basis. Long experience with this course show that about 71% of the students pass. This semester 88 students are taking Psychology 231 by correspondence. Let r be a random variable that represents the number that will pass. The Psychology Department wants a probability distribution for r.

   (a) Write a brief but complete description in which you explain why the normal approximation to the binomial would apply. Are the assumptions satisfied? Explain.

   (b) Estimate P(r $\geq$ 60).

   (c) Estimate P(60 $\leq$ r $\leq$ 70).

Chapter Test 7A

1. Write a brief but complete discussion of each of the following topics: population, parameter, sample, sampling distribution, statistical inference using sampling distributions. In each case be sure to give a complete and accurate definition of the terms. Illustrate your discussion using examples from everyday life.

2. The diameters of oranges from a Florida orchard are <u>normally distributed</u> with mean $\mu = 3.2$ inches and standard deviation $\sigma = 1.1$ inches. A packing supplier is designing special occasion presentation boxes of oranges and needs to know the average diameter for a random sample of 8 oranges. What is the probability that the mean diameter $\overline{x}$ for these oranges is
   (a) smaller than 3 inches?
   (b) longer than 4 inches?
   (c) between 3 and 4 inches?

3. The manufacturer of a new compact car claims the miles per gallon (mpg) for the gasoline consumption is mound shaped and symmetric with mean $\mu = 25.9$ mpg and standard deviation $\sigma = 9.5$ mpg. If 30 such cars are tested, what is the probability the average mpg $\overline{x}$ is
   (a) less than 23 mpg?
   (b) more than 28 mpg?
   (c) between 23 and 28 mpg?

Chapter Test 7B

1. Write a brief but complete discussion in which you cover the following topics: What is the mean $\mu_{\bar{x}}$ and standard deviation $\sigma_{\bar{x}}$ of the $\bar{x}$ distribution based on a sample of size n? Be sure to give appropriate formulas in your discussion. How do you find a standard z score corresponding to $\bar{x}$?; State the Central Limit Theorem, and the general conditions under which it can be used. Illustrate your discussion using examples from everyday life.

2. Chemists use pH to measure the acidity/alkaline nature of compounds. A large vat of mixed commercial chemicals is supposed to have a mean pH $\mu = 6.3$ with a standard deviation $\sigma = 1.9$. Assume a normal distribution for pH values. If a random sample of ten readings in the vat is taken and the mean pH $\bar{x}$ is computed, find each of the following
   (a) $P(5.2 \leq \bar{x})$
   (b) $P(\bar{x} \leq 7.1)$
   (c) $P(5.2 \leq \bar{x} \leq 7.1)$

3. Fire department response time is the length of time it takes a fire truck to arrive at the scene of the fire starting from the time the call was given to the truck. Response time for the Castle Wood Fire Department follows a mound shaped and symmetric distribution. The response time has mean $\mu = 8.8$ minutes with standard deviation $\sigma = 2.1$ minutes. If a random sample of 32 response times is taken and the mean response time $\bar{x}$ is computed, find each of the following
   (a) $P(8 \leq \bar{x})$
   (b) $P(\bar{x} \leq 9)$
   (c) $P(8 \leq \bar{x} \leq 9)$

Chapter Test 8A

1. As part of an Environmental Studies class project, students measured the circumferences of a random sample of 45 Blue Spruce trees near Brainard Lake, Colorado. The sample mean circumference was $\bar{x} = 29.8$ inches with sample standard deviation $s = 7.2$ inches. Find a 95% confidence interval for the population mean circumference of all Blue Spruce trees near this lake. Write a brief explanation of the meaning of the confidence interval in the context of this problem.

2. A random sample of 19 Rainbow Trout caught at Brainard Lake, Colorado had mean length $\bar{x} = 11.9$ inches with sample standard deviation $s = 2.8$ inches. Find a 99% confidence interval for the population mean length of all Rainbow Trout in this lake. Write a brief explanation of the meaning of the confidence interval in the context of this problem.

3. A random sample of 78 students were interviewed and 59 said they would vote for Jennifer McNamara as student body president.
   (a) Let p represent the proportion of all students at this college who will vote for Jennifer. Find a point estimate $\hat{p}$ for p.
   (b) Find a 90% confidence interval for p
   (c) What assumptions are required for the calculation of part b? Do you think these assumptions are satisfied? Explain.
   (d) How many more students should be included in the sample to be 90% sure that a point estimate $\hat{p}$ will be within a distance of 0.05 of p.

4. A random sample of 53 students were asked for the number of semester hours they are taking this semester. The sample standard deviation was found to be $s = 4.7$ semester hours. How many <u>more</u> students should be included in the sample to be 99% sure the sample mean $\bar{x}$ is within one semester hour of the population mean μ for all students at this college?

5. How long do new batteries last on a camping trip? A random sample of $n_1 = 42$ small camp flashlights were installed with brand I batteries and left on until the batteries failed. The sample mean lifetime was $\bar{x}_1 = 9.8$ hours with sample standard deviation $s_1 = 2.2$ hours. Another random sample of $n_2 = 38$ small flashlights of the same model were installed with brand II batteries and left on until the batteries failed. The sample mean of lifetimes was $\bar{x}_2 = 8.1$ hours with sample standard deviation $s_2 = 3.5$ hours.
   (a) Find a 90% confidence interval for the population difference $\mu_1 - \mu_2$ of lifetimes for these batteries.
   (b) Does the confidence interval of part a contain all positive, all negative, or both positive and negative numbers? What does this tell you about the mean life of battery I compared to battery II?

Chapter Test 8A continued

6. Two pain relief drugs are being considered. A random sample of 8 doses of the first drug showed that the average amount of time required before the drug was absorbed into the blood stream was $\bar{x}_1 = 24$ min with standard deviation $s_1 = 4$ minutes. For the second drug, a random sample of 10 doses showed the average time required for absorption was $\bar{x}_2 = 29$ min with standard deviation $s_2 = 3.9$ minutes. Assume the absorption times follow a normal distribution. Find a 90% confidence interval for the difference in average absorption time for the two drugs. Does it appear that one drug is absorbed faster than the other (at the 90% confidence level)? Explain.

7. A random sample of 83 investment portfolios managed by Kendra showed that 62 of them met the targeted annual percent growth. A random sample of 112 investment portfolios managed by Lisa showed that 87 met the targeted annual percent growth. Find a 99% confidence interval for the difference in the proportion of the portfolios meeting target goals managed by Kendra compared to those managed by Lisa. Is there a difference in the proportions at the 99% confidence level? Explain.

Chapter Test 8B

1. A random sample of 14 evenings (6 PM to 9 PM) at the O'Sullivan household showed the family received an average of $\bar{x} = 5.2$ solicitation phone calls each evening. The sample standard deviation was $s = 1.9$. Find a 95% confidence interval for the population mean number of solicitation calls this family receives each night. Write a brief explanation of the meaning of the confidence interval in the context of this problem.

2. Computer Depot is a large store that sells and repairs computers. A random sample of 110 computer repair jobs took technicians an average of $\bar{x} = 93.2$ minutes per computer. The sample standard deviation was $s = 16.9$ minutes. Find a 99% confidence interval for the population mean time μ for computer repairs. Write a brief explanation of the meaning of the confidence interval in the context of this problem.

3. A random sample of 56 credit card holders showed that 41 regularly paid their credit card bills on time.
   (a) Let p represent the population proportion of all people who regularly pay their credit card bills on time. Find a point estimate $\hat{p}$ for p.
   (b) Find a 95% confidence interval for p.
   (c) What assumptions are required for the calculations of part b? Do you think these assumptions are satisfied? Explain.
   (d) How many more credit card holders should be included in the sample to be 95% sure that a point estimate $\hat{p}$ will be within a distance of 0.05 of p?

4. At a large office supply store, the daily sales of two similar brand-name laser printers are being compared. A random sample of 16 days showed that brand I had mean daily sales $\bar{x}_1 = \$2464$ with sample standard deviation $s_1 = \$529$. A random sample of 19 days showed that brand II had mean daily sales $\bar{x}_2 = \$2285$ with sample standard deviation $s_2 = 612$. Assume sales follow an approximately normal distribution.
   (a) Find a 90% confidence interval for the population mean difference in sales $\mu_1 - \mu_2$.
   (b) Does the confidence interval of part a contain all positive, all negative, or both positive and negative numbers? What does this tell you about the mean sales of one printer compared to that of the other?

5. Allen is an appliance salesman who works on commission. A random sample of 39 days showed that the sample standard deviation for value of sales was $s = \$215$. How many more days must be included in the sample to be 95% sure the population mean μ is within $50 of the sample mean $\bar{x}$?

Chapter Test 8B continued

6. A production manager is studying the effect of overtime on different shifts. On Shift I at least half of the workers were on overtime. A random sample of 245 items from the assembly line showed that 24 were defective. Shift II had no overtime workers. A random sample of 258 items from the assembly line showed that 11 had defects.
   (a) Find a 90% confidence interval for the population proportion difference $p_1 - p_2$ of defective items for shift I versus shift II.
   (b) What assumptions are required for the calculation of part a? Do you think these assumptions are satisfied? Explain.
   (c) Does the confidence interval of part a contain all positive, all negative, or both positive and negative numbers? What does this tell you about the population proportion of defects for Shift I compared to Shift II?

7. Red Stone tires has developed a new tread which they claim reduces stopping distance on wet pavement. A random sample of 56 test drives with cars using tires with tread type I (old design) showed that the average stopping distance on wet pavement was $\bar{x}_1 = 183$ feet with sample standard deviation $s_1 = 49$ feet. A random sample of 61 test drives conducted under similar conditions, but with cars using tires with tread type II(new tread) showed that the average stopping distance was $\bar{x}_2 = 152$ feet with sample standard deviation $s_2 = 53$ feet.
   (a) Find a 90% confidence interval for the population mean difference $\mu_1 - \mu_2$ of stopping distances for the two types of tire tread.
   (b) Does the confidence interval of part a contain all positive, all negative, or both positive and negative numbers? What does this tell you about the mean stopping distance using tires with the new tread design compared to that using tires with the old tread design?

Chapter Test 9A

For each of the following problems please provide the requested information.
  (a) State the null and alternate hypotheses.
  (b) Identify the sampling distribution to be used: the standard normal or the Student's t. Find the critical value(s).
  (c) Sketch the critical region and show the critical value(s) on the sketch.
  (d) Compute the z or t value of the sample test statistic and show it's location on the sketch of part (c).
  (e) Find the P value or an interval containing the P value for the sample test statistic.
  (f) Based on your answers for parts (a) to (e), shall we reject or fail to reject the null hypothesis? Explain your conclusion in simple nontechnical terms.

1. A large furniture store has begun a new ad campaign on local television. Before the campaign, the long term average daily sales were $24,819. A random sample of 40 days during the new ad campaign gave a sample mean daily sale of $\bar{x}$ = $25,910 with sample standard deviation s = $1,917. Does this indicate that the population mean daily sales is now more than $24,819? Use a 1% level of significance.

2. A new bus route has been established between downtown Denver and Englewood, a suburb of Denver. Dan has taken the bus to work for many years. For the old bus route, he knows from long experience that the mean waiting time between buses at his stop was μ = 18.3 minutes. However, a random sample of 5 waiting times between buses using the new route had mean $\bar{x}$ = 15.1 minutes with sample standard deviation s = 6.2 minutes. Does this indicate that the population mean waiting time for the new route is different from what it used to be? Use α = 0.05.

3. The State Fish and Game Division claims that 75% of the fish in Homestead Creek are Rainbow Trout. However, the local fishing club caught (and released) 189 fish one weekend, and found that 125 were Rainbow Trout. The other fish were Brook Trout, Brown Trout, and so on. Does this indicate that the percentage of Rainbow Trout in Homestead Creek is less than 75%? Use α = 0.01.

4. A telemarketer is trying two different sales pitches to sell a carpet cleaning service. For sales pitch I, 175 people were contacted by phone and 62 of these people bought the cleaning service. For sales pitch II, 154 people were contacted by phone and 45 of these people bought the cleaning service. Does this indicate that there is any difference in the population proportions of people who will buy the cleaning service, depending on which sales pitch is used. Use α = 0.05.

Chapter Test 9A continued

5. A systems specialist has studied the work flow of clerks all doing the same inventory work. Based on this study, she designed a new work flow layout for the inventory system. To compare average production for the old and new methods, a random sample of six clerks was used. The average production rate (number of inventory items processed per shift) for each clerk was measured both before and after the new system was introduced. The results are shown below. Test the claim that the new system increases the mean number of items processed per shift. (use $\alpha = 0.05$).

Clerk	1	2	3	4	5	6
B: Old	116	108	93	88	119	111
A: New	123	114	112	82	127	122

6. How productive are employees? One way to answer this question is to study annual company profits per employee. Let x_1 represent annual profits per employee in computer stores in St. Louis. A random sample of $n_1 = 11$ computer stores gave a sample mean of $\overline{x}_1 = 25.2$ thousand dollars profit per employee with sample standard deviation $s_1 = 8.4$ thousand dollars. Another random sample of $n_2 = 9$ building supply stores in St. Louis gave a sample mean $\overline{x}_2 = 19.9$ thousand dollars per employee with sample standard deviation $s_2 = 7.6$ thousand dollars. Does this indicate that in St. Louis, computer stores tend to have higher mean profits per employee. Use $\alpha = 0.01$

7. How big are tomatoes? Some say that depends on the growing conditions. A random sample of $n_1 = 89$ organically grown tomatoes had sample mean weight $\overline{x}_1 = 3.8$ ounces with sample standard deviation $s_1 = 0.9$ ounces. Another random sample of $n_2 = 75$ tomatoes that were not organically grown had sample mean weight $\overline{x}_2 = 4.1$ ounces with sample standard deviation $s_2 = 1.5$ ounces. Does this indicate a difference either way between population mean weights of organically grown tomatoes compared to those not organically grown. Use a 5% level of significance.

Chapter Test 9B

For each of the following problems please provide the requested information.
  (a) State the null and alternate hypotheses.
  (b) Identify the sampling distribution to be used: the standard normal or the Student's t. Find the critical value(s).
  (c) Sketch the critical region and show the critical value(s) on the sketch.
  (d) Compute the z or t value of the sample test statistic and show it's location on the sketch of part (c).
  (e) Find the P value or an interval containing the P value for the sample test statistic.
  (f) Based on your answers for parts (a) to (e), shall we reject or fail to reject the null hypothesis? Explain your conclusion in simple nontechnical terms.

1. Long term experience showed that after a type of eye surgery it took a mean of $\mu = 5.3$ days recovery time in the hospital. However, a random sample of 32 patients with this type of eye surgery were recently treated as outpatients during the recovery. The sample mean recovery time was $\bar{x} = 4.2$ days with sample standard deviation $s = 1.9$ days. Does this indicate that the mean recovery time for outpatients is less than the time for those recovering in the hospital? Use a 1% level of significance.

2. Recently the national average yield on municipal bonds has been $\mu = 4.19\%$. A random sample of 16 Arizona municipal bonds gave an average yield of 5.11% with sample standard deviation $s = 1.15\%$. Does this indicate that the population mean yield for all Arizona municipal bond is greater than the national average? Use a 5% level of significance.

3. At a local four-year college, 37% of the student body are Freshmen. A random sample of 42 student names taken from the Dean's Honor List over the past several semesters showed that 17 were Freshmen. Does this indicate the population proportion of Freshmen on the Dean's Honor List is different from 37%? Use a 1% level of significance.

4. In a random sample of 62 students, 34 said they would vote for Jennifer as student body president. In another random sample of 77 students, 48 said they would vote for Kevin as student body president. Does this indicate that in the population of all students, Kevin has a higher proportion of votes? Use $\alpha = 0.05$.

5. Five members of the college track team in Denver (elevation 5,200 ft) went up to Leadville (elevation 10,152 ft) for a track meet. The times in minutes for these team members to run two miles at each location are shown in below

Team Member	1	2	3	4	5
Denver	10.7	9.1	11.4	9.7	9.2
Leadville	11.5	10.6	11.0	11.2	10.3

Assume the team members constitute a random sample of track team members. Use a 5% level of significance to test the claim that the times were longer at the higher elevation.

Chapter Test 9B continued

6. Two models of a popular pick-up truck are tested for miles per gallon (mpg) gasoline consumption. The Pacer model was tested using a random sample of $n_1 = 9$ trucks and the sample mean was $\bar{x}_1 = 27.3$ mpg with sample standard deviation $s_1 = 6.2$ mpg. The Road Runner model was tested using a random sample of $n_2 = 14$ trucks. The sample mean was $\bar{x}_2 = 22.5$ mpg with sample standard deviation $s_2 = 6.8$ mpg. Does this indicate the population mean gasoline consumption for the Pacer is higher than that of the Road Runner? Use $\alpha = 0.01$.

7. Students at the college agricultural research station are studying egg production of range free chickens compared to caged chickens. During a one week period, a random sample of $n_1 = 93$ range free hens produced an average of $\bar{x}_1 = 11.2$ eggs with sample standard deviation $s_1 = 4.4$ eggs. For the same period, another random sample of $n_2 = 87$ caged hens produced a sample average of $\bar{x}_2 = 8.5$ eggs per hen with sample standard deviation $s_2 = 5.7$. Does this indicate the population mean egg production for range free hens is higher? Use a 5% level of significance.

Chapter Test 10A

For the given data, please solve the following problems.

Taltson Lake is in the Canadian Northwest Territories. This lake has many Northern Pike. The following data was obtained by two fishermen visiting the lake. Let x = length of a Northern Pike in inches and let y = weight in pounds.

x (inches)	20	24	36	41	46
y (pounds)	2	4	12	15	20

1. Draw a scatter diagram. Using the scatter diagram (no calculations) would you estimate the linear correlation coefficient to be positive, close to zero, or negative? Explain your answer.

2. For the given data compute each of the following:
   (a) $\bar{x}$ and $\bar{y}$ (b) SS_x, SS_y and SS_{xy}
   (c) The slope b and y intercept a of the least squares line; write out the equation for the least squares line.
   (d) Graph the least squares line on your scatter plot of problem 1.

3. Compute the sample correlation coefficient r. Compute the coefficient of determination. Give a brief explanation of the meaning of the correlation coefficient and the coefficient of determination in the context of this problem.

4. Compute the standard error of estimate S_e.

5. If a 32 inch Northern Pike is caught, what is the weight in pounds as predicted by the least squares line?

6. Find a 90% confidence interval for your prediction of problem 5.

7. Using the sample correlation coefficient r computed in problem 3, test whether or not the population correlation coefficient ρ is different from zero. Use $\alpha = 0.01$. Is r significant in this problem? Explain.

Chapter Tests
Chapter Test 10B

For the given data, please solve the following problems.

Do higher paid chief executive officers (CEO's) control bigger companies? Let us study x = annual CEO salary ($ millions) and y = annual company revenue ($ billions). The following data are based on information from *Forbes* magazine and represents a sample of top US executives.

x ($ millions)	0.8	1.0	1.1	1.7	2.3
y ($ billions)	14	11	19	20	25

1. Draw a scatter diagram. Using the scatter diagram (no calculations) would you estimate the linear correlation coefficient to be positive, close to zero, or negative? Explain your answer.

2. For the given data compute each of the following:
    (a) $\bar{x}$ and $\bar{y}$ (b) SS_x, SS_y and SS_{xy}
    (c) The slope b and y intercept a of the least squares line; write out the equation for the least squares line.
    (d) Graph the least squares line on your scatter plot of problem 1.

3. Compute the sample correlation coefficient r. Compute the coefficient of determination. Give a brief explanation of the meaning of the correlation coefficient and the coefficient of determination in the context of this problem.

4. Compute the standard error of estimate S_e.

5. If a CEO has an annual salary of 1.5 million, what is his or her annual company revenue as predicted by the least squares line?

6. Find a 90% confidence interval for your prediction of problem 5.

7. Using the sample correlation coefficient r computed in problem 3, test whether or not the population correlation coefficient ρ is different from zero. Use α = 0.01. Is r significant in this problem? Explain.

Copyright © Houghton Mifflin Company. All rights reserved
T-33

Chapter Test 11A

1. Are teacher evaluations independent of grades? After midterm, a random sample of 284 students were asked to evaluate teacher performance. The students were also asked to supply their midterm grade in the course being evaluated. In this study, only students with a passing grade (A, B, or C) were included in the summary table.

Teacher Evaluation	Mid Term Grade			
	A	B	C	Row Total
Positive	53	33	18	104
Neutral	25	46	29	100
Negative	14	22	44	80
Column Total	92	101	91	284

Use a 5% level of significance to test the claim that teacher evaluations are independent of midterm grades.

2. How old are college students? The national age distribution for college students is shown below.

National Age Distribution for College Students

Age	Under 26	26-35	36-45	46-55	over 55
Percent	39%	25%	16%	12%	8%

The Western Association of Mountain Colleges took a random sample of 212 students and obtained the following sample distribution.

Sample Distribution, Western Association of Mountain Colleges

Age	Under 26	26-35	36-45	46-55	over 55
Number of Students	65	73	41	21	12

Is the sample age distribution for the Western Association of Mountain Colleges a good fit to the national distribution? Use $\alpha = 0.05$.

3. A technician tested 25 motors for toy electric trains and found that the sample standard deviation of electrical current to be $s = 4.9$ amperes.
   (a) Find a 95% confidence interval for σ, the population standard deviation of electric current in all such toy trains.
   (b) If the manufacturer specifies that $\sigma = 4.1$ amperes, does the sample data indicate that σ is larger than 4.1? Use a 1% level of significance.

4. Two methods of manufacturing large roller bearings are under study. For method I, a random sample of $n_1 = 16$ bearings had sample standard deviation of diameters $s_1 = 2.9$ mm. For method II, a random sample of 18 bearings had sample standard deviation of diameters $s_2 = 1.2$ mm. Assume the diameters follow a normal distribution. Test the claim that $\sigma_1^2 > \sigma_2^2$ using a 1% level of significance.

Chapter Test 11A continued

5. A study to determine if management style affects the number of sick leave days taken by employees in a department was conducted. Three departments with the same number of employees were studied. The management style used in one department was top down with employees having little input into decisions; in another department quality control experts made recommendations; in the last department the management gathered input informally from the employees. The total number of sick leave days taken per month by all of the employees in the department was recorded. For a random sample of 3 months, the numbers follow:

Top down management:	19 34 28
Quality teams:	16 21 15
Informal input:	15 12 14

Use one-way ANOVA to test if the mean number of sick leave days for departments managed in the various styles are different. Use $\alpha = 0.05$.

6. Will students perform better if they can choose the section of a course in which they enroll? Does the class status of the student make a difference? A researcher is studying this question. Four blocks of students are formed according to class status: Freshman, Sophomore, Junior, Senior. Each of the students must enroll in the course Spanish I. The researcher selects a random sample of 10 students from each of the blocks and allows them to enroll in the section of their choice. Another random sample of 10 students from each block are assigned a section of Spanish I. At the end of the semester, all students take the same final exam. The researcher records the scores and compares the scores for all the students participating in the study.
(a) Draw a flowchart showing the design of this experiment.
(b) Does the design fit the model for a two-way ANOVA randomized block design? Explain.

Chapter Test 11B

1. Is the choice of college major independent of grade average? A random sample of 445 students were surveyed by the Registrar's Office regarding major field of study and grade average. In this study, only students with passing grades (A, B, or C) were included in the survey. Grade averages were rounded to the nearest letter grade (e.g. 3.6 grade point average was rounded to 4.0 or A)

Major	Grade Average			
	A	B	C	Row Total
Science	38	49	63	150
Business	41	42	59	142
Humanities	32	53	68	153
Column Total	111	144	190	445

Use a 1% level of significance to test the claim that choice of major field is independent of grade average.

2. The Fish and Game Department in Wisconsin stocked a new lake with the following distribution of game fish

Initial Stocking Distribution

Fish	Pike	Trout	Perch	Bass	Glue Gill
Percent	10%	15%	20%	25%	30%

After six years a random sample of 197 fish from the lake were netted, identified, and released. The sample distribution is shown next.

Sample Distribution after Six Years

Fish	Pike	Trout	Perch	Bass	Glue Gill
Number	52	15	33	55	42

Is the sample distribution of fish in the lake after six years a good fit to the initial stocking distribution? Use a 5% level of significance.

3. An automobile service station timed the Quick Lube service for a random sample of 22 customers. The sample standard deviation of times was s = 6.8 minutes.
    (a) Find a 90% confidence interval for σ, the population standard deviation of Quick Lube times.
    (b) The service manager specifies that σ be 6.0 minutes. Does the sample data indicate that σ is different from 6.0? Use a 1% level of significance.

Chapter Test 11B continued

4. A large national chain of department stores has two basic inventories. Variation of cash flow for the two types of inventories is under study. A random sample of $n_1 = 9$ stores with inventory I had sample standard deviation of daily cash flow $s_1 = \$3,115$. Another random sample of $n_2 = 11$ stores with inventory II had sample standard deviation of daily cash flow $s_2 = \$2719$. Assume daily cash flow follows a normal distribution. Test the claim that the population variances of the two inventories are different. Use a 5% level of significance.

5. A study of depression and exercise was conducted. Three groups were used: those in a designed exercise program; a group that is sedentary; and a group of runners. A depression rating (higher scores meaning more depression) was given to the participants in each group. Small random samples from each groups provided the following data on the depression rating:

   Treatment Group: 63 58 61
   Sedentary Group: 71 64 68
   Runners: 49 52 47

   Use one-way ANOVA to test if the mean depression ratings for the three groups are different. Use $\alpha = 0.05$.

6. A study was conducted to measure sales volume of a grocery store item. The study looked at sales volume for the product placed in 3 different shelf locations: eye level, low, special display. In addition, the study looked at sales volume for the item when it was advertised in two different ways: on TV or with newspaper coupons. A two-way ANOVA test was used to determine if there was any difference in mean sales volume according to shelf location or advertising method.
   (a) List the factors and the levels of each factor for this study.
   (b) Explain what it means to have interaction between the factors. State the null and the alternate hypotheses used to test for interaction.

Chapter Test 12A

1. The management of a large retail store did a study of sales in different departments last year and this year and ranked the departments by sales volume (with lowest rank meaning highest sales). The data follow.

Department	1	2	3	4	5	6	7	8
Last year rank	1	7	4	5	3	2	6	8
This year rank	1	6	5	7	2	3	4	8

   Test the claim at the 0.01 level of significance that there is a monotone relation either way between last year's and this year's performance.

2. A workshop on harmony in the work place was given to a randomly selected group of employees. Another group did not participate in the workshop. A test measuring sensitivity to other viewpoints was given to both groups with higher scores indicating more sensitivity. The results follow.

Workshop participant	73	81	91	56	78	83	52	92
Non-workshop	85	70	74	55	90	48	75	86

   Test the claim at the 5% level of significance that there is a difference either way in the average sensitivity score for the two groups.

3. A restaurant wants to determine if a newspaper coupon giving diners a free desert increases the number of diners per night. The number of diners during the week before the coupon became effective and the number of diners each night during the week the coupon was in effect are given below.

	Mon	Tue	Wed	Thur	Fri	Sat	Sun
No Coupon	73	53	65	84	112	140	97
Coupon	82	47	70	65	120	130	160

   Use a 5% level of significance to test the hypothesis that the mean number of diners coming into the restaurant when the coupon was in effect was higher.

Chapter Test 12B

1. Professor Smith gives a midterm and final exam. For a random sample of ten students, the class rank on the two exams follow, with a lower rank number meaning a higher score.

Student	1	2	3	4	5	6	7	8	9	10
Midterm	5	3	1	7	10	2	8	4	6	9
Final	9	4	5	7	8	3	6	1	2	10

    Test the claim that there is a monotone increasing relation between the ranks of the two exams. Use a 5% level of significance.

2. A random sample of households with an income level below $30,000 were asked to record the number of hours per week someone in the household was watching T.V. A random sample of households with income level at or above $30,000 were asked to record the same information. The results follow.

Less than $30,000	100	40	35	70	80	90	15	75
$30,000 or more	85	62	41	37	45	91	30	10

    Test the claim at the 5% level of significance that there is a difference either way in the average number of hours households in the two income categories watch T.V.

3. A self confidence inventory instrument was administered to a group of students before and after a self confidence training workshop. The scores follow with a higher score indicating more self confidence.

Student	1	2	3	4	5	6	7	8	9
Before	35	42	37	45	43	47	33	37	35
After	40	38	37	43	45	46	41	40	36

    Use a 5% level of significance to test the hypothesis that the mean scores were higher after the workshop.

Solutions To Chapter Tests

Chapter Test 1A

1. (a) Length of time it took each of the Colorado residents who earned (or will earn) a Bachelor's degree to complete the degree program.
   (b) Length of time it took each of 265 recent graduates to earn a Bachelor's degree
   (c) Not necessarily.

2. (a) Nominal
   (b) Nominal
   (c) Ratio
   (d) Interval
   (e) Nominal
   (f) Ordinal
   (g) Ordinal

3. (a) Census
   (b) Experiment
   (c) Sampling
   (d) Simulation

4. *See* text.

Chapter Test 1B

1. (a) Observed book purchase (mystery or not a mystery) of all current customers of the bookstore.
   (b) Observed book purchase (mystery or not a mystery) of the 76 customers.
   (c) No.

2. (a) Nominal
   (b) Ratio
   (c) Interval
   (d) Ordinal
   (e) Ratio

3. (a) Census
   (b) Sampling
   (c) Simulation
   (d) Experiment

4. *See* text.

Chapter Test 2A

1. The outcomes are the number of dots on the face, 1 through 6. Consider single digits in the random number table. Select a starting place at random. Record the first five digits you encounter that are between (and including) 1 and 6. The first five outcomes are

    3 6 1 5 6

2.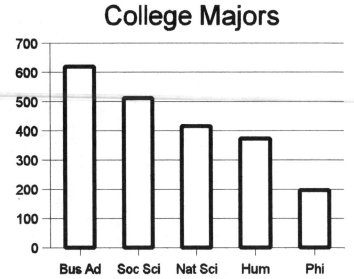

3. (a) The Class Width Is 6

    Frequency and Relative Frequency Table

Class Boundaries	Freq.	Rel. Frequency	Mid-point	Cumulative Frequency
2.5 - 8.5	10	0.3333	5.5	10
8.5 - 14.5	9	0.3000	11.5	19
14.5 - 20.5	6	0.2000	17.5	25
20.5 - 26.5	2	0.0667	23.5	27
26.5 - 32.5	3	0.1000	29.5	30

Chapter Test 2A continued

3. (b)

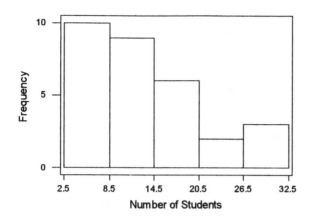

(c)

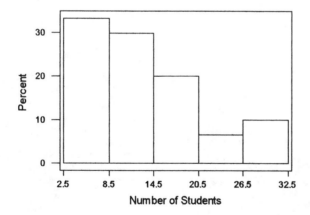

Chapter Test 2A continued

3. (d)

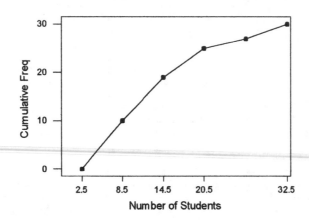

4.   0 | 5 = $5

        0 | 5 7 8 9
        1 | 2 5 6 7 9
        2 | 1 2 4
        3 | 3 5 7
        4 | 2 6 9
        5 | 1 7

5. (a) Systematic
   (b) Cluster
   (c) Stratified
   (d) Convenience
   (e) Random

6.

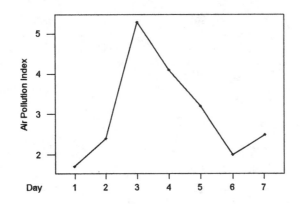

Chapter Test 2B

1. Assign each of the 736 employees a distinct number between 1 and 736. Select a starting place in the random number table at random. Use groups of three digits. Use the first 30 distinct groups of three digits that correspond to employee numbers.

    622   413   055   401   334

2.

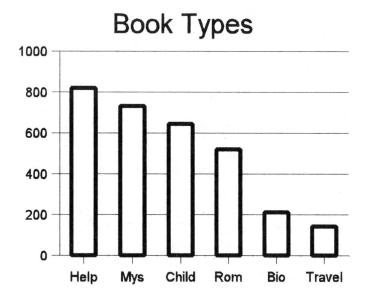

3. (a)   The Class Width Is 15

***** Frequency and Relative Frequency Table *****

Class Boundaries	Freq.	Rel. Freq.	Mid-point	Cumulative Freq.
18.5 - 33.5	3	0.1250	26	3
33.5 - 48.5	7	0.2917	41	10
48.5 - 63.5	8	0.3333	56	18
63.5 - 78.5	4	0.1667	71	22
78.5 - 93.5	2	0.0833	86	24

Chapter Test 2B continued

(b)

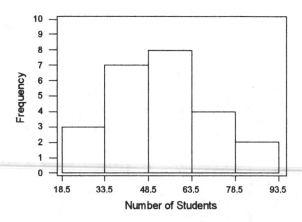

(c)

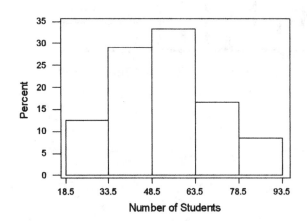

(d)

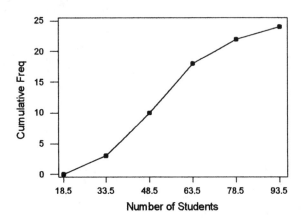

Chapter Test 2B continued

4. 2 | 1 = 21 students

    2 | 1 6 9
    3 | 3 4 6 8
    4 | 2 4 5 7
    5 | 0 1 2 6 8 9
    6 | 3 4 8

5. (a) Simple Random
    (b) Stratified
    (c) Cluster
    (d) Convenience
    (e) Systematic

6.

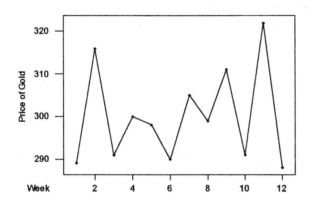

Chapter Test 3A

1. $\bar{x} = 16.61$; median = 13; mode = 12

2. (a) Range = 25   (b) $\bar{x} = 33$ (c) $s^2 = 81.67$ (d) $s = 8.37$

3. (a) CV = 21%   (b) 6.04 to 14.72

4. (a) $\bar{x} = 330$ (b) $s^2 = 130.55$ (c) $s = 11.43$

5. (a) Low value = 15; $Q_1 = 23$;   median = 33; $Q_3 = 49$; High value = 72
   (b)

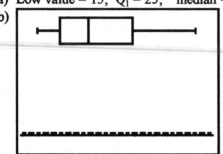

   (c) Interquartile range = 26

6. Weighted average = 85.65

Chapter Test 3B

1. $\bar{x} = 10.55$; median = 11; mode = 11

2. (a) Range = 14   (b) $\bar{x} = 12.17$ (c) $s^2 = 25.34$ (d) $s = 5.04$

3. (a) CV = 26.5%   (b) 15.3 to 49.7

4. (a) $\bar{x} = 2.35$ (b) $s^2 = 1.02$ (c) $s = 1.007$

5. (a) Low value = 27; $Q_1 = 35.5$; median = 42.5; $Q_3 = 57$; high value = 68
   (b)

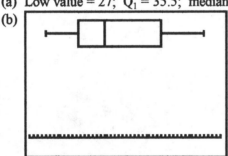

   (c) Interquartile range = 21.5

6. Weighted average = 84.75

Chapter Test 4A

1. (a) Relative frequency; 19/317 = 0.0599 or 5.99%
   (b) 1 - 0.599 = 0.9401 or about 94%
   (c) Defective, not defective; the probabilities add up to one.

2. (a) With replacement, P(Red first *and* Green second) = (3/12)(7/12) = 7/48 or 0.146
   (b) Without replacement, P(Red first *and* Green second) = (3/12)(7/11) = 21/132 or 0.159

3. P(Approval on written *and* interview) = P(written)P(interview, *given* written) =
   (0.63)(0.85) = 0.536 or about 53.6%

4. (a) 35/360   (b) 21/360   (c) 5/149   (d) 107/360
   (e) 48/149   (f) 31/360
   (g) No, P(Satisfied) = 138/360 and P(Satisfied, *given* referred) = 59/149 are not equal.

5. 8

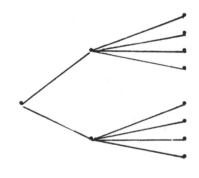

Printer   Monitor

Chapter Test 4B

1. (a) P(woman) = 4/10 = 0.4 or 40%   (b) P(man) = 6/10 = 0.6 or 60%
   (c) P(President) = 1/10 = 0.1 or 10%

2. (a) With replacement: P(Red first *and* Green second) = (6/17)(8/17) = 0.166 or 16.6%
   (b) Without replacement: P(Red first *and* Green second) = (6/17)(8/16) = 0.176 or
       17.6%

3. P(woman *and* computer science major) = P(woman)P(computer science major *given*
   woman) = (0.64)((0.12) = 0.077 or about 7.7%

4. (a) 180/417   (b) 12/115   (c) 166/417   (d) 61/417
   (e) 61/101   (f) 65/180
   (g) No, P(neutral) = 166/417 is not equal to P(neutral, given Freshman) = 61/101

5. 16·15·14·13·12 = 524,160 ways

Chapter Test 5A

1. (a) These results are rounded to three digits. $P(0) = 0.038$; $P(1) = 0.300$; $P(2) = 0.263$; $P(3) = 0.171$; $P(4) = 0.117$ $P(5) = 0.058$; $P(6) = 0.033$; $P(7) = 0.021$

    (b)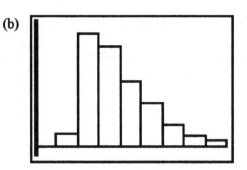

    (c) $P(2 \leq x \leq 5) = 0.609$
    (d) $P(x < 3) = 0.601$
    (e) $\mu = 2.44$
    (f) $\sigma = 1.57$

2. $n = 16$; $p = 0.65$; Success= reach summit
    (a) $P(r = 160 = 0.001$
    (b) $P(r \geq 10) = 0.840$
    (c) $P(r \leq 12) = 0.825$
    (d) $P(9 \leq r \leq 12) = 0.666$

3. (a)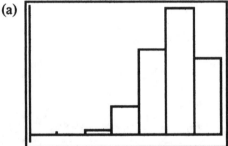

    (b) $\mu = 1.5$
    (c) $\sigma = 0.968$

4. $n = 11$ is the minimal number

5. (a) Essay; $P(n) = 0.15(0.85)^{n-1}$
    (b) $P(n = 3) = 0.108$
    (c) $P(n > 3) = 0.764$

6. (a) Essay; $\lambda = 6$; $P(r) = (e^{-6} 6^{r})/r!$
    (b) $P(r = 3) = 0.0892$
    (c) $P(r \geq 3) = 0.038$

Chapter Test 5A continued

7. (a) Essay; $\lambda = 4.5$; $P(r) = (e^{-4.5}\, 4.5^r)\, /\, r!$
    (b) $(r = 3) = 0.1687$
    (c) $P(\, r \geq 3) = 0.8264$

Chapter Test 5B

1. (a) Values are rounded to three digits. $P(1) = 0.066$; $P(2) = 0.092$; $P(3) = 0.202$
       $P(4) = 0.224$; $P(5) = 0.184$; $P(6) = 0.079$; $P(7) = 0.053$; $P(8) = 0.044$;
       $P(9) = 0.035$; $P(10) = 0.022$
    (b)

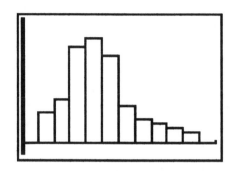

    (c) 0.641
    (d) 0.101
    (e) $\mu = 4.40$
    (f) $\sigma = 2.08$

2. Success = accept; $p = 0.30$; $n = 15$
    (a) $P(r = 15) = 0.000$ (to three digits)
    (b) $P(r \geq 8) = 0.051$
    (c) $P(\, r \leq 4) = 0.517$
    (d) $P(5 \leq r \leq 10) = 0.484$

3. (a)

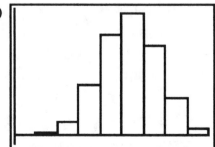

    (b) $\mu = 3.85$
    (c) $\sigma = 1.20$

4. $n = 10$ is the smallest number

Chapter Test 5B continued

5. (a) $P(n) = 0.60(0.40)^{n-1}$
   (b) $P(n = 3) = 0.0162$
   (c) $P(n \geq 3) = 0.664$

6. (a) Essay; $\lambda = 8$; $P(r) = (e^{-8} 8^r)/ r!$
   (b) 0.1396
   (c) $P(r < 3) = 0.0137$

7. (a) Essay; $\lambda = 2$; $P(r) = (e^{-2} 2^r)/ r!$
   (b) $P(r = 2) = 0.1353$
   (c) $P(r > 2) = 0.3233$

Chapter Test 6A

1. (a) $z \geq -1.8$ (b) $-1.2 \leq z \leq 0.8$ (c) $z \leq 2.2$
   (d) $x \leq 3.1$ (b) $4.1 \leq x \leq 4.5$ (c) $x \geq 4.35$

2. (a) No, look at z values
   (b) For Joel, $z = 1.51$. For John, $z = 2.77$. Relative to the district, John is a better salesman.

3. (a) $P(x \leq 5) = P(z \leq -0.77) = 0.2206$
   (b) $P(x \geq 10) = P(z \geq 1.5) = 0.0668$
   (c) $P(5 \leq x \leq 10) = P(-0.77 \leq z \leq 1.5) = 0.7126$

4. 29.92 minutes or 30 minutes

5. (a)

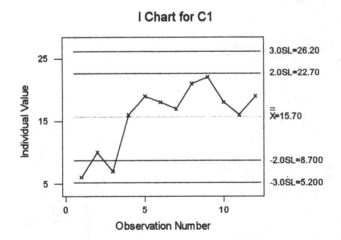

   (b) Yes; Type III for days 1 to 3; Type III; Type II for days 4 thru 12

Chapter Test 6A continued

6. (a) Essay
   (b) P(r ≤ 47) = P(x ≤ 46.5) = P(z ≤ -1.22) = 0.6112
   (c) P(47 ≤ r ≤ 55) = P(47.5 ≤ x ≤ 54.5) = P(-9.82 ≤ z ≤ 1.93) = 0.7671

Chapter Test 6B

1. (a) z ≤ 0.33   (b) 0 ≤ z ≤ 2.00   (c) z ≥ -1.33
   (d) x ≥ 2.6   (e) 6.8 ≤ x ≤ 11   (f) x ≤ 13.1

2. (a) Generator II is hotter; see the z values
   (b) z = 0.72 for Generator I; z = 2.75 for Generator II

3. (a) P(x < 6) = P(z < 1.33) = 0.918
   (b) P(x > 7) = P(z > 2.00) = 0.0228
   (c) P(6 ≤ x ≤ 7) = P(1.33 ≤ z ≤ 2.00) = 0.0690

4. z = 3 years

5. (a)

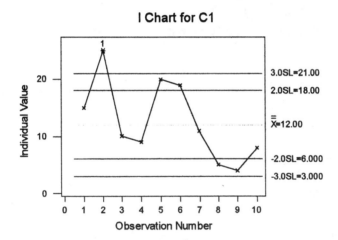

   (b) Signal I, day 2; Signal III for days 8-10.

6. (a) Essay
   (b) P(r ≥ 60) = P(x ≥ 60.5) = P(z ≥ -0.58) = 0.2810
   (c) P(60 ≤ r ≤ 70) = P(59.5 ≤ r ≤ 69.5) = P(0.70 ≤ z ≤ 1.65) = 0.1925

Chapter Test 7A

1. Essay

2. (a) $P(\bar{x} < 3) = P(z < -0.51) = 0.3050$
   (b) $P(\bar{x} > 4) = P(z > 2.06) = 0.0197$
   (c) $P(3 \leq \bar{x} \leq 4) = P(-0.51 \leq z \leq 2.06) = 0.6753$

3. (a) $P(\bar{x} < 23) = P(z < 1.67) = 0.0475$
   (b) $P(\bar{x} > 28) = P(z > 1.21) = 0.1131$
   (b) $P(23 \leq \bar{x} \leq 28) = P(1.67 \leq z \leq 1.21) = 0.0656$

Chapter Test 7B

1. Essay

2. (a) $P(5.2 \leq \bar{x}) = P(-1.83 \leq z) = 0.0336$
   (b) $P(\bar{x} \leq 7.1) = P(z \leq 1.33) = 0.0934$
   (c) $P(5.2 \leq \bar{x} \leq 7.1) = P(-1.83 \leq z \leq 1.33) = 0.8730$

3. (a) $P(8 \leq \bar{x}) = P(-2.15 \leq z) = 0.0158$
   (b) $P(\bar{x} \leq 9) = P(z \leq 0.54) = 0.2946$
   (c) $P(8 \leq \bar{x} \leq 9) = P(-2.15 \leq z \leq 0.54) = 0.6896$

Chapter Test 8A

1. 27.70 to 31.90 inches

2. 10.05 to 13.75 inches; use $t = 2.878$

3. (a) $\hat{p} = 59/78 \approx 0.756$
   (b) 0.68 to 0.84
   (c) $np > 5$ and $nq > 5$; Yes
   (d) 141 more

4. 95 more

5. (a) 0.61 to 2.79 for $\mu_1 - \mu_2$
   (b) Since the interval contains all positive numbers, it seems that battery I has a longer population mean life time.

6. $s = 3.94$; interval from -8.27 to -1.73 for $\mu_1 - \mu_2$; Since the interval contains numbers that are all negative it appears that the population mean duration of the first drug is less than that of the second.

Chapter Test 8A continued

7. $\hat{p}_1 = 62/83$ for Kendra; $\hat{p}_2 = 87/112$ for Lisa; interval from -0.19 to 0.13 for $p_1 - p_2$. Since the interval contains both positive and negative values, there is no evidence of a difference in population proportions of successful portfolios managed by Kendra compared to those managed by Lisa.

Chapter Test 8B

1. 4.10 to 6.30 calls; use t = 2.160

2. 89.0 to 97.4 minutes

3. (a) $\hat{p} = 41/56$ or 0.732
   (b) 0.62 to 0.85
   (c) np and nq are both greater than 5
   (d) 246 more

4. (a) s = 575.75; -151.6 to 509.62 dollars for $\mu_1 - \mu_2$
   (b) Since the interval contains both positive and negative values, it does not appear that the population mean daily sales of the printers differ.

5. 33 more

6. (a) $\hat{p}_1 = 0.98$; $\hat{p}_2 = 0.043$; 0.018 to 0.093 for $p_1 - p_2$
   (b) $n_1p_1, n_1q_1, n_2p_2, n_2q_2$ are all greater than 5
   (c) The interval contains all positive values and shows that at the 90% confidence level, the population proportion of defects is greater on Shift I.

7. 15.5 to 46.5 for $\mu_1 - \mu_2$; the interval contains values that are all positive. At the 90% confidence level, the population mean stopping distance of the old thread design is greater than that for the new.

Chapter Test 9A

1. H_0: $\mu = 24,819$; H_1: $\mu > 24,819$; $z_0 = 2.33$; $\bar{x} = 25,910$ corresponds to z = 3.60; P value = 0.0002; Reject H_0. There is evidence that the population mean daily sales is greater.

2. H_0: $\mu = 18.3$; H_1: $\mu \neq 18.3$; $t_0 = \pm 2.776$; d.f. = 4; $\bar{x} = 15.1$ corresponds to t = -1.154; P value > 0.250; Do not reject H_0. There is not enough evidence to condlude that the waiting times are different.

3. H_0: p = 0.75; H_1: p < 0.75; $z_0 = -2.33$; $\hat{p} = 125/189 = 0.661$ corresponds to z = -2.81; P value = 0.0024; Reject H_0. There is evidence that the proportion of Rainbow Trout is less than 75%.

Chapter Test 9A continued

4. H_0: $p_1 = p_2$; H_1: $p_1 \neq p_2$; $z_0 = \pm 1.96$; $\hat{p}_1 = 62/175 = 0.354$; $\hat{p}_2 = 45/154 = 0292$; $\hat{p}_1 - \hat{p}_2 = 0.059$ corresponds to $z = 1.20$; P value $= 0.2304$; Do not reject H_0. There is no evidence of a difference in population proportions of success between the two different sales pitches.

5. H_0: $\mu_d = 0$; H_1: $\mu_d < 0$; d.f. $= 5$; $t_0 = 2.015$; $s_d = 8.17$; $\bar{d} = -7.5$ corresponds to $t = -2.263$;
   $0.025 <$ P value < 0.050; Reject H_0. There is evidence that the new process increases the mean number of items processed per shift.

6. H_0: $\mu_1 = \mu_2$; H_1: $\mu_1 > \mu_2$; d.f. $= 18$; $t_0 = 2.522$; $\bar{x}_1 - \bar{x}_2 = 5.3$ corresponds to $t = 1.464$; $0.075 <$ P value < 0.10; Do not reject H_0. There is not sufficient evidence to show that the population mean profit per employee is higher in St. Louis

7. H_0: $\mu_1 = \mu_2$; H_1: $\mu_1 \neq \mu_2$; $z_0 = \pm 1.96$; $\bar{x}_1 - \bar{x}_2 = -0.3$ corresponds to $z = -1.517$; P value $= 0.1291$; Do not reject H_0. There is not enough evidence to say that the mean weight of tomatoes grown organically is different than that of other tomatoes.

Chapter Test 9B

1. H_0: $\mu = 5.3$; H_1: $\mu < 5.3$; $z_0 = -2.33$; $\bar{x} = 4.2$ corresponds to $z = -3.28$; P value $= 0.0005$; Reject H_0. There is evidence that the average recovery time is less as an outpatient.

2. H_0: $\mu = 4.19$; H_1: $\mu > 4.19$; d.f. $= 15$; $t_0 = 1.753$; $\bar{x} = 5.11$ corresponds to $t = 3.200$; P value < 0.005; Reject H_0; There is evidence that the average yield of Arizona municipal bonds is higher than the national average yield.

3. H_0: $p = 0.37$; H_1: $p \neq 0.37$; $z_0 = \pm 1.96$; $\hat{p} = 17/42 = 0.405$ corresponds to $z = 0.47$; P value $= 0.6408$; Do not reject H_0. There is not sufficient evidence to conclude that the proportion of Freshmen on the Dean's List is different from that in the college.

4. H_0: $p_1 = p_2$; H_1: $p_1 < p_2$; $z_0 = -1.645$; $\hat{p}_1 = 0.540 = $ proportion in favor of Jennifer; $\hat{p}_2 = 0.623 = $ proportion in favor of Kevin; $\hat{p}_1 - \hat{p}_2 = -0.084$ corresponds to $z = -1.00$; P value $= 0.1586$; Do not reject H_0. There is not enough evidence to say that the proportion who plan to vote for Kevin is higher.

5. H_0: $\mu_d = 0$; H_1: $\mu_d < 0$; d.f. $= 4$; $t_0 = -2.132$; $s_d = 0.78$; $\bar{d} = -0.9$ corresponds to $t = -2.566$; $0.025 <$ P value < 0.050; Reject H_0. There is evidence that the average time for runners at higher elevation is longer.

6. H_0: $\mu_1 = \mu_2$; H_1: $\mu_1 > \mu_2$; d.f. $= 21$; $t_0 = 2.518$; $\bar{x}_1 - \bar{x}_2 = 4.8$ corresponds to $t = 1.708$; $s = 6.578$; $0.05 <$ P value < 0.075; Do not reject H_0. There is not enough evidence that the average mileage for the Pacer is greater.

Chapter Test 9B

7. H_0: $\mu_1 = \mu_2$; H_1: $\mu_1 > \mu_2$; $z_0 = 1.645$; $\bar{x}_1 - \bar{x}_2 = 2.7$ corresponds to $z = 3.54$; P value $= 0.0002$. Reject H_0. The evidence indicates that the average egg production of range free chickens is higher.

Chapter Test 10A

1. Length of Northern Pike x versus Weight y

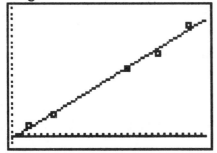

   The linear correlation appears positive.

2. (a) $\bar{x} = 33.4$; $\bar{y} = 10.6$ (b) $SS_x = 491.2$; $SS_y = 227.2$; $SS_{xy} = 332.8$
   (c) $b = 0.6775$; $a = -12.029$; $y = 0.678x - 12.03$
   (d) *See* line on the graph of Problem 1.

3. $r = 0.996$; $r^2 = 0.992$; 99.2% of the variance of y is due to variance in x and the least squares model.

4. $S_e = 0.757$

5. For x = 32 inches, y = 9.65 pounds

6. $7.70 \le y \le 11.61$ pounds

7. H_0: $\rho = 0$; H_0: $\rho \neq 0$; $r_0 = 0.96$; sample $r = 0.996$; There is evidence that the population correlation coefficient is not equal to 0. r is significant.

Chapter Test 10B

1. (a) CEO salary x, versus Company Revenue y

   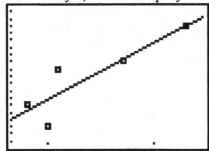

   (b) The correlation appears to be positive.

2. (a) $\bar{x} = 1.38$; $\bar{y} = 17.8$ (b) $SS_x = 1.508$; $SS_y = 118.8$; $SS_{xy} = 11.78$
   (c) $b = 7.811678$; $a = 7.019883$; $y = 7.811x + 7.02$
   (d) *See* the line on the graph in Problem .

3. $r = 0.880$; $r^2 = 0.775$; 77.5% of the variance in y can be explained by the variance in x and the least squares model.

4. $S_e = 2.98767$

5. For a CEO salary of 1.5 million dollars, we predict an annual company revenue of 18.74 billion dollars.

6. $11.01 \leq u \leq 26.47$ billion dollars for company revenue.

7. H_0: $\rho = 0$; H_0: $\rho \neq 0$; $r_0 = 0.96$; sample statistic $r = 0.880$; reject H_0. The sample statistic is not significant. We do not have evidence of correlation between CEO salary and company revenue at the 1% level of significance.

Chapter Test 11A

1. H_0: Teacher evaluations are independent of Midterm grades; H_1: Teacher evaluations are not independent of Midterm grades; $\chi^2 = 43.68$; $\chi^2_{0.05} = 9.49$; Reject H_0. There is evidence to say that at the 5% level of significance teacher evaluations are not independent of midterm grades.

2. H_0: The distribution of ages of college students is the same in the Western Association as in the nation. H_1: The distribution of ages of college students is different in the Western Association that it is in the nation. $\chi^2 = 15.03$; $\chi^2_{0.05} = 9.49$; Reject H_0. There is evidence that the distribution of ages is different.

3. (a) $3.83 \leq \sigma \leq 6.82$
   (b) H_0: $\sigma = 4.1$; H_1: $\sigma > 4.1$; $\chi^2_{0.01} = 42.98$; $\chi^2 = 34.28$; Do not reject H_0, There is not enough evidence to reject $\sigma = 4.1$

Chapter Test 11A

4. H_0: $\sigma_1^2 = \sigma_2^2$; H_1: $\sigma_1^2 > \sigma_2^2$; d.f.$_N$ = 15; d.f.$_D$ = 17; critical value is 3.31; sample F = 5.84; Reject H_0; There is evidence that the variance of diameters in method I is greater.

5. H_0: all the means are equal; H_1: not all the means are equal; SS_{BET} = 48.67; SS_W = 375.33; SS_{TOT} = 424; d.f.$_{BET}$ = 2; d.f.$_W$ = 6; d.f.$_{TOT}$ = 8; MS_{BET} = 24.33; MS_W = 62.56; $F_{0.05}$ = 5.14; Sample F = 0.34; Do not reject H_0. The means do not seem to be different among the groups.

6. (a)

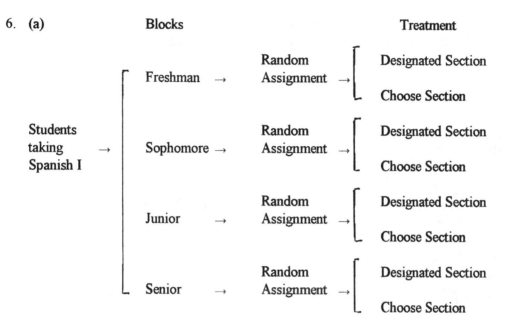

(b) Yes

Chapter Test 11B

1. H_0: The choice of college major is independent of grade average; H_1: The choice of college major is not independent of grade average. $\chi^2_{0.05}$ = 13.28; χ^2 = 2.64; Do not reject H_0. There is not enough evidence to conclude that college major is dependent of grade average.

2. H_0: The distribution of fish fits the initial stocking distribution. H_1: The distribution of fish after six years does not fit the initial stocking distribution. $\chi^2_{0.01}$ = 9.49; χ^2 =66.78; Reject H_0. There is enough evidence to conclude that the fish distribution has changed.

3. (a) $5.45 \leq \sigma \leq 9.15$
   (b) H_0: $\sigma = 6$; H_1: $\sigma \neq 6$; critical values are 8.03 and 41.40; χ^2 = 27.97; Do not reject H_0. There is not enough evidence to conclude that the standard deviation is different from 6.

Chapter Test 11B continued

4. H_0: $\sigma_1^2 = \sigma_2^2$; H_1: $\sigma_1^2 \neq \sigma_2^2$; $d.f._N = 8$; $d.f._D = 10$; use $\alpha = 0.025$ to find the critical value of 3.85; sample $F = 1.312$; Do not reject H_0; The variances of the two inventories do not seem to be different.

5. H_0: all the means are equal; H_1: not all the means are equal; $SS_{BET} = 14.89$; $SS_W = 548.67$; $SS_{TOT} = 563.56$; $d.f._{BET} = 2$; $d.f._W = 6$; $d.f._{TOT} = 8$; $MS_{BET} = 7.44$; $MS_W = 91.44$; $F_{0.05} = 5.14$; Sample $F = 0.08$; Do not reject H_0. The means do not seem to be different among the groups.

6. (a) Factor 1 shelf location with 3 level: eye level, low, special
    Factor 2 advertisement with 2 levels: TV, newspaper
    (b) There is interaction if sales volume for levels in factor 1 differ according to levels of factor 2. H_0: There is no interaction between the factors; H_1: There is some interaction

Chapter Test 12A

1. H_0: $\rho_S = 0$; H_1: $\rho_S \neq 0$; critical value $= 0.881$; Since the sample test statistic $r_S = 0.857$ does not fall in the critical region, we fail to reject H_0. There does not seem to be a monotone relation.

2. H_0: Workshop makes no difference; H_1: Workshop makes a difference; $z_0 = \pm 1.96$; $\mu_R = 68$; $\sigma_R = 9.522$; R(workshop participants) $= 73$ corresponds to $z = 0.525$; Since the sample test statistic does not fall in the critical region, we fail to reject H_0. The workshop does not seem to make a difference.

3. Let $\mu_1 = $ average number before coupon and $\mu_2 = $ average number when coupon is in effect; H_0: $\mu_1 = \mu_2$; H_1: $\mu_1 < \mu_2$; $z_0 = -1.645$; The sample proportion of plus signs r $= 3/7$ corresponds to $z = -0.377$. Since the sample test statistic does not fall in the critical region, we fail to reject H_0. There appears to be no difference in the average number of diners.

Chapter Test 12B

1. H_0: $\rho_S = 0$; H_1: $\rho_S > 0$; critical value = 0.564; Since the sample test statistic $r_S = 0.588$ falls in the critical region, we reject H_0. There seems to be a monotone increasing relation between midterm and final exam scores.

2. H_0: Income level makes no difference; H_1: Income level makes a difference; $z_0 = \pm 1.96$; $\mu_R = 68$; $\sigma_R = 9.522$; R(less than \$30,000) = 75 corresponds to $z = 0.735$; Since the sample test statistic does not fall in the critical region, we fail to reject H_0. Income level does not seem to make a difference.

3. Let μ_1 = average before workshop and μ_2 = average after workshop; H_0: $\mu_1 = \mu_2$; H_1: $\mu_1 < \mu_2$; $z_0 = -1.645$; The sample proportion of plus signs r = 0.375 corresponds to $z = -0.707$; Since the sample test statistic does not fall in the critical region, we fail to reject H_0. The average confidence score seems to be the same.

Part III

Answers and Key Steps to Solutions

of Even Numbered Problems

Chapter 1

Section 1.1
2. Population: gasoline mileage for *all* new 1997 cars; sample: the gasoline mileage of the 35 new 1997 cars tested.
4. Population: shelf life of *all* Healthy Crunch granola bars; sample: shelf life of the 10 bars tested.
6. Form B
8. (a) Ordinal   (b) Ratio   (c) Nominal   (d) Interval   (e) Ratio   (f) Nominal
10. (a) Sampling   (b) Simulation   (c) Census   (d) Experiment

Chapter Problems
2. Population: opinions of *all* listeners; sample: opinions of fifteen callers
4. Name, Social Security Number, color of hair, address, phone, place of birth, college major are all nominal; letter grade on test is ordinal; year of birth is interval; height and distance from home to college are ratio.

Chapter 2

Section 2.1
2. Answers vary. Use groups of 3 digits
4. Answers vary. Use groups of 3 digits
6. Answers vary. Use single digits with odd corresponding to heads and even to tails.
8. Answers vary. Use groups of 3 digits
10. Assign students distinct numbers and use the random-number table.
12. In all cases, assign distinct numbers to the items, and use a random-number table.
14. Answers vary. Use single digits with even corresponding to true and odd to false.
16. (a) Stratified   (b) Simple random   (c) Cluster   (d) Systematic
     (e) Convenience

Section 2.2
2. Number of U.S. Bills Printed and Delivered in 1989
(in millions of bills)

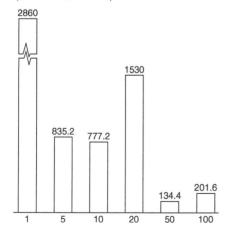

Section 2.2 continued

4. (a) Causes for Business Failure—Pareto Chart

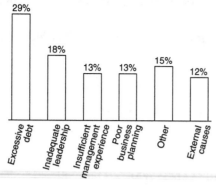

   (b) No; debt.

6. Meals We Are Most Likely to Eat in a Fast-Food Restaurant

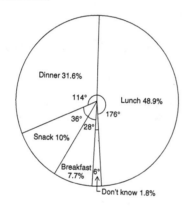

8. Age Distribution of Professors

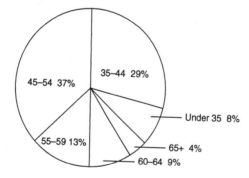

Section 2.2 Continued

10. Driving Problems—Pareto Chart

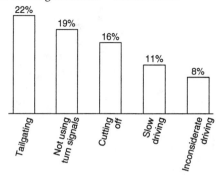

12. Social Security Tax as a Percentage of Wage, 1940–2000

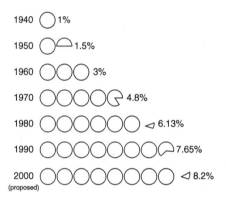

14. Time Plot: Percent Change in Employment (Anchorage, Alaska)

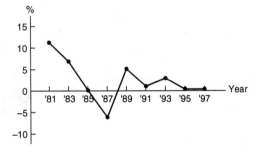

Section 2.2 continued

16. Changes in Boys' Height with Age

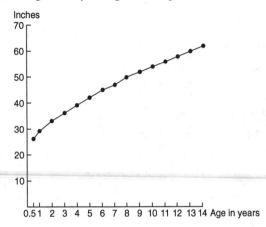

Section 2.3

2. (a) class width = 10

   (b)

   Percent Difficult Ski Terrain

Class Limits	Class Boundaries	Midpoint	Frequency	Relative Frequency	Cumulative Frequency
20–29	19.5–29.5	24.5	3	0.0857	3
30–39	29.5–39.5	34.5	6	0.1714	9
40–49	39.5–49.5	44.5	13	0.3714	22
50–59	49.5–59.5	54.5	9	0.2571	31
60–69	59.5–69.5	64.5	4	0.1143	35

   (c–e) Percent Difficult Ski Terrain—Histogram, Frequency Polygon, Relative-Frequency Histogram

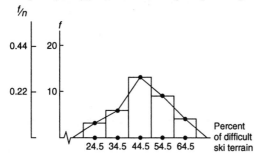

   (f) Ogive for Percent Difficult Ski Terrain

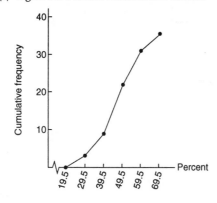

Section 2.3 continued

4. (a) Class width = 15 (round up from 14 to 15)
    (b)

Fast-Food Franchise Fees

Class Limits	Class Boundaries	Midpoint	Frequency	Relative Frequency	Cumulative Frequency
5–19	4.5–19.5	12	21	0.3333	21
20–34	19.5–34.5	27	35	0.5556	56
35–49	34.5–49.5	42	5	0.0794	61
50–64	49.5–64.5	57	1	0.0159	62
65–79	64.5–79.5	72	1	0.0159	63

(c–e) Fees for Fast-Food Franchises—Histogram, Frequency Polygon, Relative-Frequency Histogram

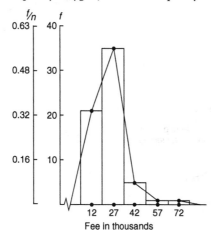

(f) Ogive for Fees

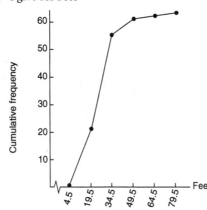

Section 2.3 continued

6. (a) Class width = 5 (round up from 4 to 5)
   (b)

Length of Time on a Major League Team

Class Limits	Class Boundaries	Midpoint	Frequency	Relative Frequency	Cumulative Frequency
1–5	0.5–5.5	3	15	0.3261	15
6–10	5.5–10.5	8	16	0.3478	31
11–15	10.5–15.5	13	9	0.1957	40
16–20	15.5–20.5	18	5	0.1087	45
21–25	20.5–25.5	23	1	0.0217	46

(c–e) Length of Time on a Major League Team—Histogram, Frequency Polygon, and Relative-Frequency Histogram

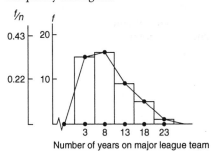

Number of years on major league team

(f) Ogive for Time on Major League Team

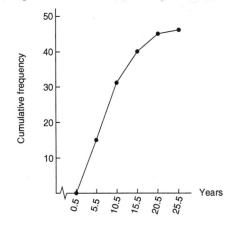

Section 2.3 continued

8. (a) Class width = 6

   (b)

Words of Three Syllables or More

Class Limits	Class Boundaries	Midpoint	Frequency	Relative Frequency	Cumulative Frequency
0–5	−0.5–5.5	2.5	13	0.24	13
6–11	5.5–11.5	8.5	15	0.27	28
12–17	11.5–17.5	14.5	11	0.20	39
18–23	17.5–23.5	20.5	3	0.05	42
24–29	23.5–29.5	26.5	6	0.11	48
30–35	29.5–35.5	32.5	4	0.07	52
36–41	35.5–41.5	38.5	2	0.04	54
42–47	41.5–47.5	44.5	1	0.02	55

(c–e) Words of Three Syllables or More—Histogram, Frequency Polygon, Relative-Frequency Histogram

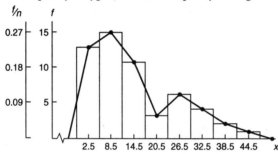

(f) Ogive for Words of Three Syllables or More

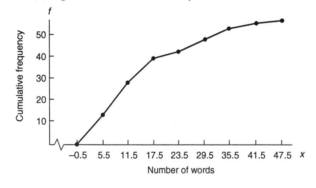

Section 2.3 continued

10. (a) Class midpoints: 34.5; 44.5; 54.5; 64.5; 74.5; 84.5

    (b)

Ages of Representatives—Frequency Polygon

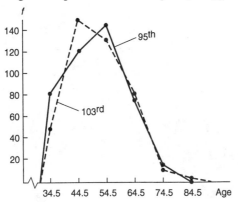

    (c) 103rd house has more members in the forties, sixties, and eighties

12.

    (a) **Miami Dolphins (*cw* = 21)**

Class Limits	Midpoint	Frequency
175–195	185	13
196–216	206	7
217–237	227	19
238–258	248	8
259–279	269	11
280–300	290	12

Weights of Football Players—Miami Dolphins

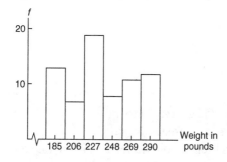

San Diego Chargers (*cw* = 32)

Class Limits	Midpoint	Frequency
119–150	134.5	1
151–182	166.5	4
183–214	198.5	27
215–246	230.5	15
247–278	262.5	14
279–310	294.5	11

Section 2.3 continued

12. continued

Weights of Football Players—San Diego Chargers

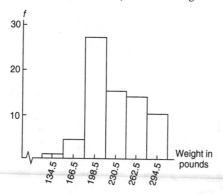

(b) Scales are different; use same class limits

14. (a)

Ogive for Average Cost per Day

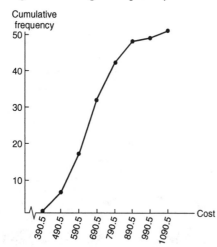

(b) 32

16. (a) Uniform; skewed right; bimodal; bimodal; symmetrical

(b) Appeal to income group which occurs most freuently.

(c) Somewhat questionable, since information is volunteered and cannot be checked.

18. (a) Class width = 0.043

Baseball Batting Averages

Class Limits	Boundaries	Midpoint	Frequency
0.107–0.149	0.1065–0.1495	0.128	3
0.150–0.192	0.1495–0.1925	0.171	4
0.193–0.235	0.1925–0.2355	0.214	3
0.236–0.278	0.2355–0.2785	0.257	10
0.279–0.321	0.2785–0.3215	0.3	6

Section 2.3 continued
18. continued

(b, c) Baseball Batting Averages—Histogram

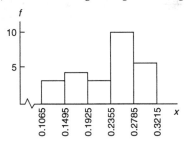

Section 2.4
2.

Percent of Wetlands Lost

4	0 = 40%
0	9
1	
2	0 3 4 7 7 8
3	0 1 3 5 5 5 6 7 8 8 9
4	2 2 6 6 6 8 9 9
5	0 0 0 2 2 4 6 6 9 9
6	0 7
7	2 3 4
8	1 5 7 7 9
9	0 1

Section 2.4 continued

4.

Number of Hospitals per State

0	8 = 8 hospitals		
0	8	15	
1	1 2 5 6 9	16	2
2	1 7 7	17	5
3	5 7 8	18	
4	1 2 7	19	3
5	1 2 3 9	20	9
6	1 6 8	21	
7	1	22	7
8	8	23	1 6
9	0 2 6 8		
10	1 2 7	42	1
11	3 3 7 9	43	
12	2 3 9	44	0
13	3 3 6		
14	8		

Texas and California have the highest number of hospitals.

6.

(a) **First Round Scores**

6	5 = score of 65
6•	5 6 7 7
7*	0 1 1 1 1 1 1 1 1 1 1 2 2 2 3 3 3 3 4 4 4
7•	5 5 5 5 5 5 5

(b) **Fourth Round Scores**

6	8 = score of 68
6•	8 9 9 9 9 9
7*	0 0 0 0 1 1 1 1 1 1 1 1 2 2 2 2 2 2 2 3 3 3 3 3 3 4 4 4

(c) Scores are lower in the fourth round. In the first round both the low and high scores were more extreme than in the fourth round.

Section 2.4 continued

8.

Radio Brightness

0	90 = 09.0 units
0	90 90 94 95 95 95 98
1	05 10 15 15 15 25 25 35 36 37 65 65 65
2	00 00 80
3	
4	40 40 40 40
5	
6	70

The measurement 67.0 is unusually bright.

10.

Milligrams of Carbon Monoxide

1	5 = 1.5 mg CO		
1	5	11	
2		12	3 6
3		13	0 6 9
4	9	14	4 9
5	4	15	0 4 9
6		16	3 6
7		17	5
8	5	18	5
9	0 5		
10	0 2 2 6	23	5

Chapter 2 Review Problems

2. (a) 322 to 322PPM   (b) 345 to 355 PPM

4. (a)

Age of DUI Arrests

1	6 = 16 years
1	6 8
2	0 1 1 2 2 2 3 4 4 5 6 6 6 7 7 7 9
3	0 0 1 1 2 3 4 4 5 5 6 7 8 9
4	0 0 1 3 5 6 7 7 9 9
5	1 3 5 6 8
6	3 4

(b) Class width = 7

Age Distribution of DUI Arrests

Class Limits	Class Boundaries	Midpoint	Frequency	Relative Frequency	Cumulative Frequency
16–22	15.5–22.5	19	8	0.16	8
23–29	22.5–29.5	26	11	0.22	19
30–36	29.5–36.5	33	11	0.22	30
37–43	36.5–43.5	40	7	0.14	37
44–50	43.5–50.5	47	6	0.12	43
51–57	50.5–57.5	54	4	0.08	47
58–64	57.5–64.5	61	3	0.06	50

Chapter 2 Review Problems continued

 4.

(c) Age Distribution of DUI Arrests—Histogram

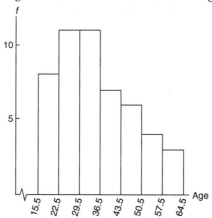

(d) Ogive for Age of DUI Arrests

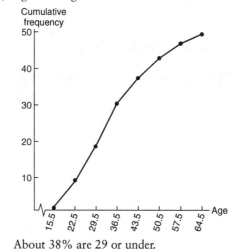

About 38% are 29 or under.

Chapter 2 Review Problems continued

6.

(a) Distribution of Civil Justice Caseloads Involving Business—Pareto Chart

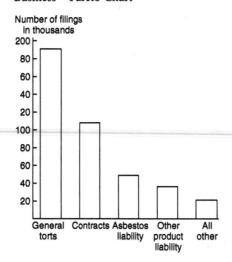

(b) Distribution of Civil Justice Caseloads Involving Business—Pie Chart

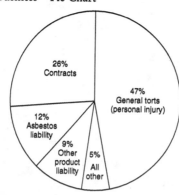

Chapter 2 Review Problems continued

8. (a) Skewed left

   (b)

   High School Grade Point Average of Students Who Graduated from College

Class Boundaries	Relative Frequency	Cumulative Frequency
0.75–1.25	1%	1%
1.25–1.75	1%	2%
1.75–2.25	2%	4%
2.25–2.75	8%	12%
2.75–3.25	17%	29%
3.25–3.75	27%	56%
3.75–4.25	44%	100%

   (c) 29% less than 3.25; 56% less than 3.75

10. (a) Use a random-number table to obtain 30 distinct three-digit numbers between 001 and 950.

    (b) Distribution of Makes of Cars in Parking Lot—Pareto Chart

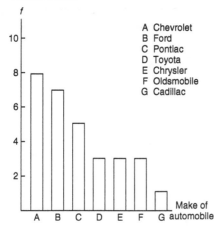

    Distribution of Makes of Cars in Parking Lot—Pie Chart

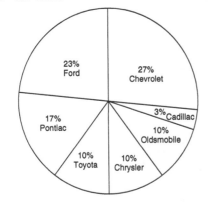

Chapter 2 Review Problems continued

10. (c) Class width = 5

Model Years for Thirty Cars in Student Parking Lot

Class Limits	Class Boundaries	Midpoint	Frequency
1976–1980	1975.5–1980.5	78	1
1981–1985	1980.5–1985.5	83	2
1986–1990	1985.5–1990.5	88	8
1991–1995	1990.5–1995.5	93	10
1996–2000	1995.5–2000.5	98	9

Model Year of Car—Histogram

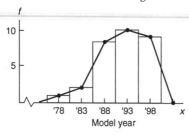

Histogram is skewed left

(d) In part b the data are nominal so we cannot make a histogram; in part c the data are interval. We can total the number of cars in each interval and make a circle graph showing the percentage in each model year interval.

12. (a) Cluster   (b) Convenience   (c) Systematic   (d) Simple random
    (e) Stratified

Chapter 3

Section 3.1

2. Mean = 45.17; median = 46.5; mode = 46
4. $\bar{x} \approx 6.2$; median = 6; mode = 7
6. (a) $\bar{x} \approx 26.3$; median = 25.5; mode = 25
    (b) Median, answers are very close.
8. (a) Mean = 7.4; median = 8; mode = 8
    (b) Mean = 14.69; median = 8; mode = 8
    (c) Generally, the mean changes the most.
10. (a) Mean = 3.09; median = 2.9; mode = 3.3
    (b) Mean = 11.025; median = 6.75; mode = 6.6
    (c) The mean is very sensitive to extreme values.
12. (a) Mean, median, and mode if it exists
    (b) Mode if it exists
    (c) Mean, median, and mode if it exists
14. Discussion question
16. Answers may vary. Here are some examples.
    (a) 1 2 2 2 3   (b) 1 2 2 2 13   (c) 1 1 5 5 5
    (d) 1 2 4 5 5   (e) -2 -1 0 1 2

Section 3.2

2. (a) Range = 60.8; $\bar{x} \approx 54.1$
   (b) $s^2 \approx 377.78$; $s \approx 19.44$
   (c) CV $\approx 35.9\%$; s is 35.9% of $\bar{x}$
4. (a) Range = 7.3; $\bar{x} \approx 9.1$; $s^2 = 8.88$; $s \approx 2.98$; CV $\approx 32.7\%$
   (b) Range = 1.9; $\bar{x} \approx 26.1$; $s^2 \approx 19.79$; $s \approx 4.45$; CV $\approx 17.0\%$
   (c) More relatively consistent productivity at a higher average level.
6. (a) Ralph: range = 4; Gloria: range = 7.2
   (b) Ralph: $\bar{x} = 21.6$; $s = 1.53$; CV = 7.1%; Gloria: $\bar{x} = 21.4$; $s = 3.22$; CV = 15%
   (c) Ralph got more consistent milage, and his CV is lower.
8. (a) Results round to answer given
   (b) 386 to 1074
   (c) 213 to 1246
10. 1.6% to 6.4%
12. 0.2% to 0.6%
14. The mean for Arms artifacts is twice that of Stable artifacts
16. Since CV = $(s/\bar{x})100$, then $s = CV(\bar{x})/100$ and $s = 0.033$

Section 3.3

2. $\bar{x} \approx 16.1$; $s^2 \approx 119.9$; $s \approx 10.95$
4. (a) $\bar{x} \approx 20.35$ (b) $s \approx 3.703$ (c) CV $\approx 18.2\%$
6. (a) $\bar{x} \approx 27$; $s \approx 20.01$; CV $\approx 74.1\%$
   (b) $\bar{x} \approx 18$; $s \approx 13.29$; CV $\approx 73.8\%$
8. $\bar{x} \approx 4.6$; $s^2 \approx 6.8$; $s \approx 2.6$
10. $\bar{x} \approx 37.7$; $s^2 \approx 279.6$; $s \approx 16.7$
12. $\bar{x} \approx 6.16$; $s^2 \approx 11.45$; $s \approx 3.38$
14. 87.75; since the weights are the same, we could have computed the mean of the four scores directly.
16. (a) 8.09   (b) 8.18; this rating is higher.
18. (a) 8.53   (b) 9.38   (c) WEighted average is sensitive to extreme values
    (d) Answers vary

Section 3.4

2. 75th percentile
4. Timothy

Section 3.4 continued

6. Low = 275; Q_1 = 319; median = 335; Q_3 = 355; high = 393; IQR = 36

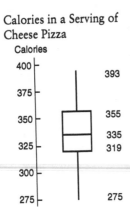

Calories in a Serving of
Cheese Pizza

8. (a) Low = 3; Q_1 = 16; median = 23; Q_3 = 30; high = 72; IQR = 14

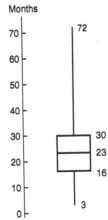

Clerical Staff Length of
Employment (months)

(b) Compare to problem 7

Section 3.4 continued

10. Low = 25; $Q_1 = 35$; median = 40; $Q_3 = 49$; high = 85; IQR = 14

Charge for Call During
Peak Time from Home

Cents per minute

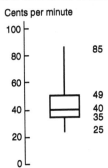

12. (a) Yes; data above median is more spread. (b) Coca Cola   (c) Coca Cola
    (d) McDonald's   (e) Disney   (f) Coca Cola

14. (a) Low 14.9; $Q_1 = 17.1$;   median = 18.9; Q_3 20.6; high = 27; IQR = 3.5

Per Capita Disposable Income
by State (thousands of dollars)

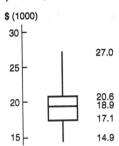

(b) Essay.

Chapter 3 Review Problems

2. (a) $\bar{x} = 4.53$; median = 4.05; mode = 1.9
   (b) $s = 2.46$; CV = 54.4%; range = 6.7
4. (a) 85.77   (b) 82.17
6. $\bar{x} \approx 44.4$; $s \approx 11.8$
8. Low = 45; $Q_1 = 71$; median = 80; $Q_3 = 84$; high = 109; IQR = 13

Glucose Blood Level After
12-Hour Fast (mg/100 ml)

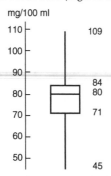

10. $\bar{x} = 4.7$; $s^2 = 3.94$; $s = 1.98$
12. (a) Use calculator   (b) -0.51 to 11.69

Chapter 4

Section 4.1

2. Answers vary.
4. (a) Cannot be negative.   (b) Must be ≤ 1
   (c) 120% = 1.20 is too large.   (d) Yes
6. Answers vary ; Probability as relative frequency
8. (a) P(never) = 0.20; P(less than 1) = 0.24; P(1 to 2) = 0.21; P(more than 2) = 0.35
   (b) Yes (within round-off error)
10. (a) Yes, assuming no one wars both glasses and contact lenses
    (b) P(no glasses) = 0.44; P(no contacts) = 0.964
    (c) Neither = complement of wears corrective lenses = 0.404
12. (a) 0.81   (b) 0.19   (c) Germinate or not germinate   (d) No
14. (a) 0.46   (b) 0.43   (c) 0.20   (d) 0.57

Section 4.2

2. (a) 30% (b) 40% (c) 80%; Yes; The probability of blue and the probability of brown differ between the types of M&M candies.
4. (a) 111/288 (b) 81/288 (c) 237/288 (d) 159/288 (e) 18/288
6. (a) Yes (b) 1/36 (c) 1/36 (d) 1/18
8. (a) 1/6 (b) 1/18 (c) Yes; 2/9
10. (a) No (b) 0.006 (c) 0.006 (d) 0.012
12. (a) Yes (b) 0.0059 (c) 0.0059 (d) 0.0118
14. (a) 0.0175 (b) 0.21 (c) 0.133; 0.012 (d) 0.07; 0.12
16. (a) 0.354 or 35.4% (b) 0.969 or 96.9%
18. P($600 or more) = 0.283; P($199 or less) = 0.371
20. (a) 110/130 (b) 20/130 (c) 50/70 (d) 20/70 (e) 110/200 (f) 20/200
22. (a) 291/2008 (b) 77/452 (c) 826/2008 (d) 131/373
    (e) 41/157 (f) 53/157 (g) 420/452 (h) 332/373
    (i) No: P(15+ yr) = 535/200;8 ≠ P(15+ yr, <u>given</u> East) = 118/452
24. (a) 0.70 (b) 0.595 (c) 0.90 (d) 0.73
    (f) In the case of *and*, we are looking at the sample space of all students, both male and female. In the case of *given*, we are restricting our sample space to females only.
26. (a) 0.27; 0.23; 0.73; 0.77 (b) 0.70; 0.95 (c) 0.189; 0.69
    (d) 0.31 (e) 0.69 (f) 0.189 (g) 0.31

Section 4.3

2.
  (a) Outcomes of Tossing
      a Coin and Throwing a Die

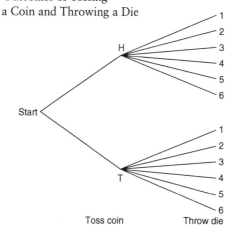

  (b) 2 (c) 1/6

Section 4.3 continued

4.

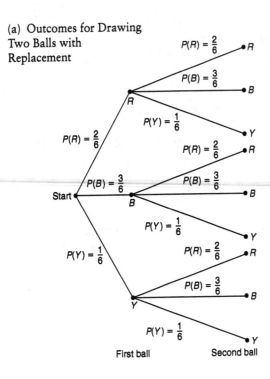

(a) Outcomes for Drawing
Two Balls with
Replacement

$P(R) = \frac{2}{6}$

$P(B) = \frac{3}{6}$

$P(Y) = \frac{1}{6}$

$P(R) = \frac{2}{6}$

$P(B) = \frac{3}{6}$

$P(Y) = \frac{1}{6}$

$P(R) = \frac{2}{6}$

$P(B) = \frac{3}{6}$

$P(Y) = \frac{1}{6}$

$P(R) = \frac{2}{6}$

$P(B) = \frac{3}{6}$

$P(Y) = \frac{1}{6}$

Start

R

B

Y

First ball   Second ball

(b) 1/9; 1/6; 1/18; 1/6; 1/4; 1/12; 1/18; 1/12; 1/36

Section 4.3 continued

6.
  (a) Outcomes of Three Multiple-Choice Questions

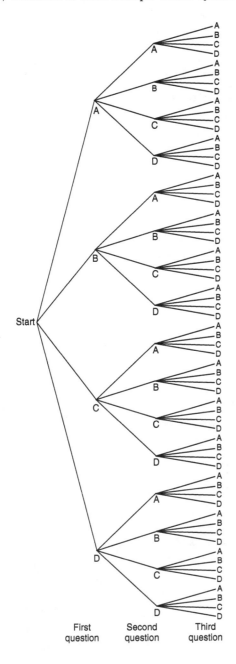

   (b) 1/64
8. 24
10. (a) 36   (b) 9   (c) 0.25
12. 60
14. 336
16. 362,800

Section 4.3 continued

18. 56
20. 1
22. $P_{10,3} = 720$
24. $P_{6,6} = 720$
26. $C_{10,3} = 120$
28. (a) $C_{12,5} = 792$ (b) 0.001   (c) 21; 0.027
30. (a) $C_{42,6} = 5,245,786$ (b) 0.00000019   (c) 0.0000019

Chapter 4 Review Problems

2. 20%; 59%; 11.8%
4. (a) Sample space: 1H, 2H, 3H, 4H, 5H, 6H, 1T, 2T, 3T, 4T, 5T, 6T
   (b) Yes   (c) 0.167
6. (a) 0.470; 0.390; 0.140   (b) 0.840; 0.040
   (c) 0.100; 0.240   (d) 0.420; 0.060
   (e) 0.860; yes   (f) No
8. 0.693
10. (a) 42   (b) 21   (c) 6   (d) 1
12.

Ways to Satisfy Literature, Social Science, and Philosophy Requirements

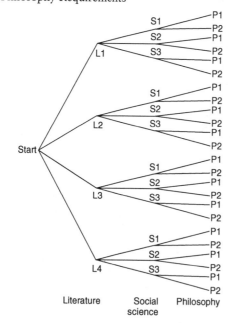

14. 64
16. $P_{3,3} = 6$

Chapter 5
Section 5.1

2. (a) continuous (b) continuous (c) discrete (d) continuous (e) discrete

4. (a) Yes; events are distinct, probabilities total to 1

   b) Age of Promotion-Sensitive Shoppers

   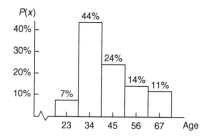

   (c) 42.58   (d) 12.31

6. (a)

   Sizes of Families

   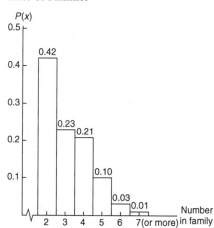

   (b) 0.42   (c) 0.35   (d) $\mu = 3.12$ (e) $\sigma = 1.20$

8. (a) Yes; outcomes are distinct and probabilities total to 1

   (b) Age of Nurses

   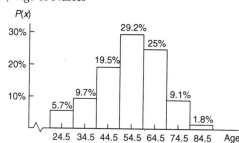

   (c) 35.9%   (d) 53.76   (e) 13.66

Section 5.1 continued

10. (a) 0.763 (b) 0.367 (c) 0.016 (d) $\mu = 1.25$ (e) $\sigma = 0.968$
12. (a) 0.0021; 0.9979 (b) $4.20; less; $25.80
14. (a) 0.00756; $378
    (b) $412.50; $448; $482.50; $517.50; $2238.50 total
    (c) $2938.50
    (d) $2761.50

Section 5.2

2. $n = 10$; $p = 0.2$
   (a) 0.000 (to three digits) (b) 0.107
   (c) 0.892; 0.893; They should be equal, but because of round off error they differ
       slightly
   (d) $P(r \geq 5) = 0.033$
4. $n = 7$; $p = 0.10$
   (a) 0.478 (b) 0.522 (c) 0.974
6. $n = 20$; $p = 0.10$
   (a) 0.878 (b) 0.323 (c) 0.122 (d) 0.677
8. $n = 11$; $p = 0.35$
   (a) 0.000 to three digits (b) 0.332 (c) 0.050 (d) 0.426
10. (a) 0.683 (b) 0.633 (c) 0.166 (d) 0.001 (e) 0.003
12. (a) 0.002 (b) 0.410 (c) 0.028
14. $n = 20$; $p = 0.70$
    (a) 0.036 (b) 0.000 to three3 digits (c) 0.000 to three digits
    (d) 0.762
16. (a) 0.740; 0.473; 0.135
    (b) 0.961; 0.117; 0.493
18. (a) 0.764 (b) 0.149 (c) 0.851 (d) 1 to three digits
    (e) 0.000 to three digits
20. (a) $p = 0.20$; 0.056 (b) $p = 0.25$; 0.629 (c) $p = 0.45$; 0.849
    (d) $p = 0.30$; 0.000 to three digits (e) $p = 0.80$; 0.168
22. (a) 0.00152 (b) 0.00001 to five places (c) 0.00153

Section 5.3

2. (a) II (b) I (c) III (d) IV
   (e) More symmetrical when p is close to 0.5; skewed left when p is close to 1; skewed
       right when p is close to 0.

Section 5.3 continued

4.   (a) Binomial Distribution for Number of Defective
Syringes

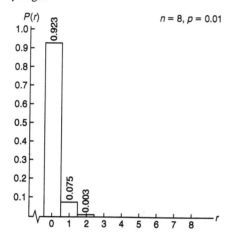

  (b) $\mu = 0.08$ (c) 0.998   (d) $\sigma = 0.281$

6.

  (a) Binomial Distribution for Number of Automobile
Damage Claims by People Under Age 25

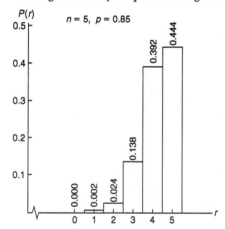

  (b) $\mu = 4.25$; $\sigma = 0.798$;   expected number is about 4.

8.

  (a) Drivers Who Tailgate

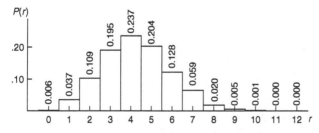

(b) 4.2   (c) 1.65

Section 5.3 continued

10.

(a, b) Binomial Distribution for the Number of Hot Spots That Are Forest Fires

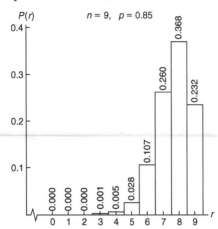

(c) $\mu = 7.65$ (d) $\sigma = 1.071$ (e) 4

12. 12 calls

14. (a) 0.972   (b) $\mu = 2.8$; $\sigma \approx 1.30$ (c) 12 cars

16. (a) 4   (b) 2.6

18. (a) 0.001   (b) 0.890   (c) 20 calls

20. (a) 6   (b) 4.95 or about 5

22. (a) 0.008   (b) 0.028   (c) 0.836   (d) $\mu = 2.7$; $\sigma = 1.21$
    (e) 10 professors

Section 5.4

2. (a) $P(n) = (0.57)(0.43)^{n-1}$ (b) 0.2451   (c) 0.1054   (d) 0.0795

4. (a) $P(n) = (0.80)(0.20)^{n-1}$ (b) 0.8; 0.16; 0.032   (c) 0.008
   (d) $P(n) = (0.04)(0.96)^{n-1}$; 0.04; 0.0384; 0.0369

6. (a) $P(n) = (0.036)(0.964)^{n-1}$ (b) 0.03345; 0.0311; 0.0241   (c) 0.8636

8. (a) $\gamma = 7.5$ per 50 liter; $P(r) = e^{-7.5}(7.5)^r / r!$
   (b) 0.0156; 0.0389; 0.0729   (c) 0.9797 (d) 0.0203

10. (a) $\gamma = 3.7$ per 11 hours (rounded to the nearest 10th)
    (b) 0.9753   (c) 0.7164   (d) 0.247

12. (a) $\gamma = 7.0$ per 50 ft   (b) 0.9704   (c) $\gamma = 2.8$ per 20 ft; 0.5305
    (d) $\gamma = 0.3$ per 2 ft; 0.0037   (e) Discussion

14. (a) Essay   (b) $\gamma = 1.00$ per 22 yr; 0.6321   (c) 0.3679
    (d) $\gamma = 2.27$ per 50 yr; 0.8967   (e) 0.1033

16. (a) $\gamma = 2.1$ (b) 0.1224   (c) 0.6203   (d) 0.6203

18. (a) $\gamma = 6.0$ (b) 0.9380; 0.5543
    (c) $\gamma = 13.0$; 0.0000 to four places; 0.9893; 0.6468

20. (a) $\gamma \approx 2.7$ (b) 0.0672   (c) 0.7513   (d) 0.2858

Chapter 5 Review Problems

2. (a) 0.378; 0.179; 0.247; 0.189; 0.007
    (b) 5.28 yrs; 4.88 yrs
4. (a) 0.924   (b) 0.595
6. 160 flights; 5.66 flights
8. 77.9; 1.97
10. 15 bonds
12. (a) small ratio or accidents per number of flights
    (b) $\lambda = 2.4$ per 100,000; 0.0907
    (c) $\lambda = 4.8$ per 200,000; 0.7058
14. $\lambda = 0.1$; 0.9048; 0.0905
16. (a) 0.83   (b) 0.165

Chapter 6

Section 6.1

2. $\mu = 16$; $\mu + \sigma = 18$; $\sigma = 2$
4.

(a) Normal Curve

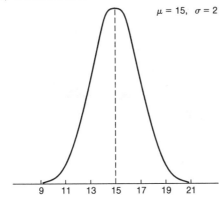

$\mu = 15$, $\sigma = 2$

(b) Normal Curve

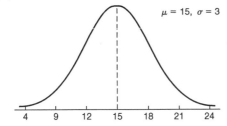

$\mu = 15$, $\sigma = 3$

Section 6.1 continued

4.
    (c) Normal Curve

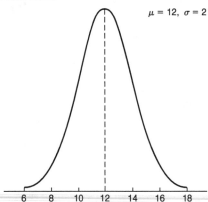

    (d) Normal Curve

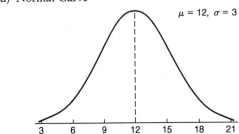

    (e) No, the values of μ and σ are independent.

6. (a) 50%   (b) 95.4%   (c) 0.15%

8. (a) 95.4% or 954 chicks   (b) 68.2% or 682 chicks

    (c) 50% or 500 chicks   (d) 99.7% or 997 chicks

10. (a) 0.159   (b) 0.841   (c) About 136 cups

12.

    (a) Visitors Treated Each Day by YPMS (First Period)

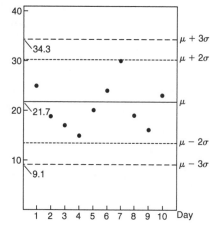

    In control.

Section 6.1 continued

12.   (b) Visitors Treated Each Day by YPMS
    (Second Period)

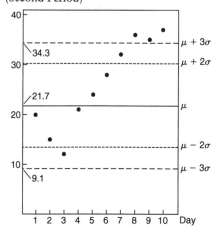

Out-of-control signals I and III are present.

14.   (a) Number of Rooms Rented (First Period)

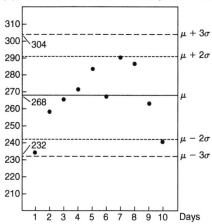

In control.

(b) Number of Rooms Rented (Second Period)

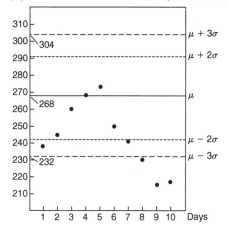

Out-of-control signals I and III are present.

Section 6.2

2. (a) -0.24   (b) 0.84   (c) 0.96   (d) 0.56
    (e) -0.72   (f) 0.16
4. (a) $z < 0.65$ (b) $-1.91 < z$ (c) $1.12 < z < 1.81$
    (d) $17.9 < x$ (e) $x < 32.7$ (f) $18.6 < x < 33.4$
    (g) $z = -3.04$; this is very small.   (h) $z = 3$ for very large fawn
6. (a) $0.86 < z$ (b) $z < -0.86$ (c) $-2.29 < z < -1.71$
    (d) $x < 9513$ (e) $11,332 < x$ (f) $7938 < x < 9688$
    (g) Unusually low, since $z = -2.9$.
8. (a) Site 1, $z = -0.63$; site 2, $z = 2.80$ (b) Site 2 is more unusual.
10. 0.4982
12. 0.4732
14. 0.8980
16. 0.0628
18. 0.4641
20. 0.5000
22. 0.4404
24. 0.3192
26. 0.9850
28. 0.7642
30. 0.2054
32. 0.4911
34. 0.0409
36. 0.0718
38. 0.8369
40. 0.5000
42. 0.0150
44. 0.0158
46. 0.9332
48. 0.9993

Section 6.3

2. 0.8914
4. 0.0471
6. 0.1693
8. 0.0918
10. About 0.9999
12. 1.96
14. -0.95
16. -1.63
18. 1.645
20. ±1.96
22. (a) 0.9664   (b) 0.9664   (c) 0.9328   (d) 0.0336
24. (a) SAT of 628; ACT of 26
    (b) SAT of 584; ACT of about 23
    (c) SAT of 475; ACT of 16.5

Section 6.3 continued

26. (a) 0.8036   (b) 0.0228   (c) 0.1736
28. (a) 00228   (b) 0.2420   (c) 0.2061
30. (a) About 21.2%   (b) 22 months
32. (a) σ ≈ 2.5 yrs   (b) 0.1151   (c) 0.0548   (d) About 9.9 years
34. (a) σ ≈ 5.25 oz   (b) 0.0228   (c) 0.0526   (d) 0.9246
    (e) About 17.8 oz
36. (a) 0.0548   (b) 0.0359   (c) 0.9093   (d) 4.9 hrs after doors open
    (e) 2.9 hrs after doors open   (f) Might be different because of work schedules
38. (a) 0.3132   (b) 0.3067

Section 6.4

2. (a) 0.0132   (b) 0.1131   (c) 0.9744   (d) 0.2514
4. (a) 0.8023   (b) 0.9671   (c) 0.8156   (d) Yes
6. (a) 0.7881   (b) 0.9936   (c) 0.7817   (d) Yes
8. (a) 0.9370   (b) 0.9842   (c) 0.9633   (d) 0.7157
10. (a) 0.0571   (b) 0.0021   (c) 0.9408   (d) Yes
12. (a) 0.9999   (b) 0.9992   (c) 0.9993
14. (a) 0.1587   (b) 0.8686   (c) 0.3175   (d) 0.6246
16. Essay

Chapter 6 Review Problems

2. (a) 0.2734   (b) 0.4332   (c) 0.0371   (d) 0.0594
    (e) 0.1660   (f) 0.9778
4. (a) 0.7967   (b) 0.9938   (c) 0.2865
6. -2.33
8. ±2.58   (Note that we do not interpolate because the areas are changing very slowly this far into the tails of the curve.)
10. (a) 336.5   (b) 261.25   (c) 0.9544
12. (a) 0.5000   (b) 4260 hours

Chapter 6 Review Problems continued

14.

(a) Hydraulic Pressure in Main Cylinder of Landing
Gear of Airplanes (psi)—First Data Set

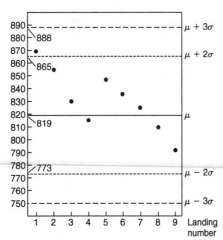

In control.

(b) Hydraulic Pressure in Main Cylinder of Landing
Gear of Airplanes (psi)—Second Data Set

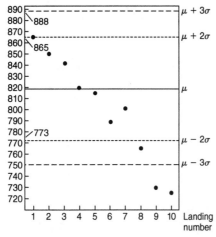

Out of control with signals I and III present.

16. (a) 0.6103   (b) 0.9573   (c) 0.6872
18. (a) 0.6700   (b) 0.3300   (c) 0.3963

Chapter 7

Section 7.1

2. See Section 2.1
4. A numerical, descriptive measure of a sample. Examples: $\bar{x}$, s, s^2, $\hat{p}$, and so forth.
6. A probability distribution for a sample statistic.
8. Relative frequencies can be thought of as a mesure or estimate of the likelihood of a certain statistic falling within the class bounds.

Section 7.2

Note: Answers may differ slightly depending on how many digits are carried in the standard deviation.

2. (a) Mean 100; standard deviation 5.33; 0.4332
   (b) Mean 100; standard deviation 4.36; 0.4664
   (c) The standard deviation is smaller for part b.
4. (a) No, sample size is too small.
   (b) Normal with mean 72 and standrd deviation 2; 0.6687
6. (a) 0.2586   (b) 0.6826
8. (a) 0.0110   (b) 0.0006   (c) Less than 0.0001
   (d) The probabilities decreased as n increases. It would be an extremely rare event for a person to have two or three tests below 3500 purely by chance.
10. (a) Normal with mean 16 and standard deviation 0.3651.
    (b) 0.4969   (c) 0.0031
12. (a) x is a mean of a sample of n = 100 stocks. By Central Limit Theorem the x distribution is approximately normal.
    (b) 0.9210   (c) 0.9823
    (d) Yes, probaiblity increases as the standard deviation decreases
    (e) 0.0007; Event is unlikely.
14. 0.9488   (b) 0.9914   (c) Yes, send a second guard
16. (a) Approximately normal with mean $20 and standard deviation $0.7.
    (b) 0.9958   (c) 0.2282   (d) Yes
18. (a) 0.8238   (b) 0.9808   (c) 0.8046
20. (a) 0.8212; 0.0104   (b) 0.9993; no

Chapter 7 Review Problems

2. All the $\bar{x}$ distributions will be normal with mean 15. The standard deviations will be 3/2, 3/4, and 3/10, respectively.
4. (a) 0.2743   (b) 0.0287   (c) The standard deviation of $\bar{x}$ is smaller
6. 08664
8. (a) Miami: 0.7019; Fairbanks: 0.7157
   (b) Miami: 0.9177; Fairbanks; 0.9332
   (c) Miami: 0.9983; Fairbanks: 0.9992

Chapter 8

Section 8.1

2. (a) 14.05 to 17.37 (b) 14.63 to 16.53
    (c) 15.07 to 16.11 (d) Lengths decrease as n increases.

4. (a) The mean and standard deviation round to the values given.
    (b) 4.28 thousand to 5.92 thousand (c) Yes (d) Yes
    (e) 3.84 thousand to 6.36 thousand; 3 thousand is below; 6.5 thousand is above

6. (a) 9.20 to 10.28 (b) 1058 to 1182
8. (a) 92.5 to 101.5 (b) Go up.
10. (a) 5.43 to 5.67 (b) 1.97 to 2.09
12. (a) 18.99 to 20.01 (b) 22.04 to 23.56
14. (a) $94.34 to $103.16 (b) No. It is high.
16. 9.87 to 11.3
18. (a) The mean and standard deviation round to the values given.
    (b) 27.03 yr to 28.57 yr (c) Yes
    (d) 26.25 yr to 29.35 yr; 33 years is still above the upper limit.

Section 8.2

2. 5.841
4. 2.201
6. (a) Use calculator. (b) 2.59 to 4.57
8. (a) Use calculator. (b) $72.55 to $94.95
10. (a) Use calculator (b) 4.95 to 6.31
12. $\bar{x} = 244.5$; s = 21.73; 217.6 to 271.4
14. Use calculator; 88.7 thousand to 125.1 thousand
16. (a) Use calculator; 3.67% to 5.73%
    (b) Use calculator, 2.55% to 3.81%
18. (a) Use calculator; 6.7 days to 7.9 days
    (b) Use calculator; 57.7% to 66.9%

Section 8.3

2. (a) $\hat{p} = 0.5491$ (b) 0.49 to 0.61 (c) Yes
4. (a) $\hat{p} = 0.6081$ (b) 0.57 to 0.65 (c) Yes
6. (a) $\hat{p} = 0.1600$ (b) 0.12 to 0.20 (c) Yes
8. (a) $\hat{p} = 0.7600$ (b) 0.75 to 0.77 (c) Yes
10. (a) $\hat{p} = 0.2900$ (b) 0.23 to 0.35 (c) Yes
12. (a) $\hat{p} = 0.5900$ (b) 0.55 to 0.63
14. (a) $\hat{p} = 0.5213$ (b) 0.47 to 0.58
    (c) A recent study shows that about 52% of medical doctors have a solo practice with margin of error 5.4 percentage points.

Section 8.3 continued

16. (a) $\hat{p} = 0.2727$ (b) 0.25 to 0.30
    (c) A recent study shows that 27.3% of all shoppers stock up on a real supermarket bargain with margin of error 2.8 percent.
18. $\hat{p} = 0.19$; $E = 0.03$; 0.16 to 0.22

Section 8.4

2. 866
4. 76 total or 35 more
6. (a) 456   (b) 257
8. (a) 106   (b) 66
10. Estimate a mean; 109 or 72 more
12. 68 total or 28 more
14. 326 total or 291 more
16. (a) 97   (b) 52 total or 14 more
18. (a) 271   (b) 214
20. 1068

Section 8.5

2. (a) $E = 1.0675$; interval from 0.13 to 2.27
   (b) Yes, average age of baseball players appears to be greater than that of basketball players..
   (c) $E = 1.5270$; interval from -0.33 to 2.73;   No difference apparent at this confidence level.
4. (a) Use calculator.
   (b) $s = 1.5133$; interval from -1.3 to 0.9
   (c) No difference.
6. (a) $\hat{p}_1 = 0.3520$; $\hat{p}_2 = 0.3800$; $\hat{\sigma} = 0.0320$; -0.08 to 0.02
   (b) No difference.
8. (a) $E = 7.5111$; interval from 2.93 to 17.95
   (b) Yes, mothers' mean score is higher than the fathers'
10. (a) Use calculator.
    (b) $s = 8.6836$; interval from 3.89 to 14.05
    (c) Yes. Average weight of grey wolves from Chihuahua is greater than those in Durango.
12. (a) $\hat{p}_1 = 0.6161$; $\hat{p}_2 = 0.1857$; $\hat{\sigma} = 0.05650$; 0.28 to 0.58
    (b) Greater proportion of artifacts seems to be unidentified at higher elevations.
14. (a) $\hat{p}_1 = 0.8196$; $\hat{p}_2 = 0.2243$; $\hat{\sigma} = 0.0297$; 0.57 to 0.63
    (b) It seems that the treatment does make a difference.
16. (a) $\hat{p}_1 = 0.5696$; 0.53 to 0.61   (b) $\hat{p}_2 = 0.3354$; 0.30 to 037
    (c) $\hat{\sigma} = 0.0282$; 0.18 to 0.29; separated nesting boxes yield more wood ducks.
18. (a) $n = (z_c/E)^2(\sigma_1^2 + \sigma_2^2)$
    (b) $n = 358.04$ or 359 players from each sport.
    (c) $n = 365.9$ or 366 players from each sport.

Chapter 8 Review Problems

2. Interval for a mean, large sample; 730 to 770
4. Sample size for mean: 102
6. Interval for mean, small sample
   (a) Use calculator.   (b) 14.27 to 17.33
8. Sample size for proportion; 9,589
10. Sample size for proportion; $n = 258$ total or 91 more.
12. Difference of means, large samples
   (a) 1.91% to 5.29%
   (b) Yes. It appears that profit as a percentage of stockholder equity is higher for retail stocks.
14. Difference of means, small samples
   (a) $s = 1.0511$; 1.35 to 2.85
   (b) Yes. It appears that the average litter size of wolf pups in Ontario is greater.
16. Difference of two proportions
   (a) $\hat{p}_1 = 0.533$; $\hat{p}_2 = 0.5435$; -0.2027 to 0.1823
   (b) No. At the 90% confidence level we do not detect any differences in the proportions.

Chapter 9

Section 9.1

2. The alternate hypothesis.
4. No
6. (a) H_0: $\mu = \$20.04$ (b) H_1: $\mu > \$20.04$
   (c) H_1: $\mu < \$20.04$ (d) H_1: $\mu \neq \$20.04$ (e) right; left; both sides.
8. (a) H_0: $\mu = 8.7$ sec.   (b) H_1: $\mu > 8.7$ sec.   (c) H_1: $\mu < 8.7$ sec.
   (d) right; left.

Section 9.2

2. H_0: $\mu = 38$ hours; H_1: $\mu < 38$ hours; left-tailed; normal; $z_0 = -2.33$; sample $z = -2.86$; Reject H_0; The assembly time is less; statistically significant.
4. H_0: $\mu = 3218$; H_1: $\mu > 3218$; right-tailed; normal; $z_0 = 2.33$; sample $z = 3.93$; Reject H_0; Average number of people entering store each day has increased; statistically significant.
6. H_0: $\mu = 17.2$; H_1: $\mu < 17.2$; left-tailed; normal; $z_0 = -1.645$; sample $z = -1.87$; Reject H_0; There is evidence that this hay has lower protein content; statistically significant.
8. H_0: $\mu = 159$ feet; H_1: $\mu < 159$ feet; left-tailed; normal; $z_0 = -2.33$; sample $z = -3.14$; Reject H_0; Mean stopping time is less; statistically significant.
10. H_0: $\mu = 233.72$ mph; H_1: $\mu \neq 233.72$ mph; two-tailed; normal; $z_0 = \pm2.58$; sample $z = -2.39$; Do not reject H_0; There is not enough evidence to conclude that the lap time average speed is different from that of Buddy Lazier; not statistically significant.
12. H_0: $\mu = 0.25$ gal; H_1: $\mu > 0.25$ gal; right-tailed; normal; $z_0 = 1.645$; sample $z = 3.00$; Reject H_0. The supplier's claim is too low; statistically significant.

Section 9.2 continued

14. (a) one-tailed test   (b) two-tailed test   (c) Yes
16. (a) For $\alpha = 0.01$, $c = 0.99$; interval from 20.28 to 23.72; Hypothesized $\mu = 20$ is not in the interval; Reject H_0.
    (b) H_0: $\mu = 20$; H_1: $\mu \neq 20$; $z_0 = \pm 2.58$; sample $z = 3.00$; Reject H_0; results the same.

Section 9.3

2. Sample $z = -0.57$; P value $= 0.2843$; No, not significant.
4. Sample $z = 0.44$; P value $= 0.660$; No, not significant.
6. H_0: $\mu = 12$; H_1: $\mu \neq 12$; sample $z = 0.66$; P value $= 0.5092$; No, not significant
8. H_0: $\mu = \$61,400$; H_1: $\mu < \$61,400$; sample $z = -1.92$; P value $= 0.0274$; Yes, significant at 5% level.
10. (a) Yes.
    (b) Right-tailed; sample $z = 1.87$; P value $= 0.0307$; Not significant at the 1% level; Significant at the 5% level; Minimum α for rejection is 0.0307 or 3%.
    (c) Two-tailed; sample $z = 1.87$; P value $= 0.0614$; Not significant at either 1% or 5% level; Minimum α for rejection is 0.0614 or 6% level.

Section 9.4

2. $t_0 = 2.681$
4. $t_0 = -1.740$
6. $t_0 = 2.467$
8. (a) Answers used in part b
    (b) H_0: $\mu = 14$; H_1: $\mu > 14$; right-tailed; $t_0 = 2.718$; sample $t = 5.53$; P value < 0.005; Reject H_0. Population average HC for this patient is higher than 14.
10. (a) Answers used in part b
    (b) H_0: $\mu = 8.8$; H_1: $\mu \neq 8.8$; two-tailed; $t_0 = \pm 2.160$; $t = -1.34$; $0.200 < $ P value < 0.250; Do not reject H_0; We cannot conclude the catch is different from 8.8 fish/day.
12. (a) Answers used in part b.
    (b) H_0: $\mu = \$38$; H_1: $\mu < 38$; left-tailed; $t_0 = -1.753$; sample $t = -2.55$; $0.010 < $ P value < 0.025; Reject H_0; Population mean car rental is less than $38.
14. (a) Answers used in part b.
    (b) H_0: $\mu = 40$; H_1: $\mu \neq 40$: two-tailed; $t_0 = \pm 2.571$; sample $t = -2.04$; $0.050 < $ P value < 0.100; Do not reject H_0; Population average heart rate for the lion is not significantly different.

Section 9.5

2. H_0: $p = 0.221$; H_1: $p < 0.221$; left-tailed; normal; $z_0 = -1.645$; $\hat{p} = 0.1658$ with $z = -1.85$; P value = 0.0322; Reject H_0.

4. H_0: $p = 0.73$; H_1: $p > 0.73$; right-tailed; normal; $z_0 = 1.645$; $\hat{p} = 0.8049$ with $z = 1.08$; P value = 0.1401; Do not reject H_0.

6. H_0: $p = 0.75$; H_1: $p \neq 0.75$; two-tailed; normal; $z_0 = \pm 1.96$; $\hat{p} = 0.7711$ with $z = 0.44$; P value = 0.66; Do not reject H_0.

8. H_0: $p = 0.214$; H_1: $p > 0.214$; right-tailed; normal $z_0 = 2.33$; $\hat{p} = 0.2759$ with $z = 3.35$; P value = 0.0004; Reject H_0.

10. H_0: $p = 0.80$; H_1: $p < 0.80$; left-tailed; normal; $z_0 = -1.645$; $\hat{p} = 0.7652$ with $z = -0.93$; P value = 0.1762; Do not reject H_0.

12. H_0: $p = 0.12$; H_1: $p < 0.12$; left-tailed; normal; $z_0 = -2.33$; $\hat{p} = 0.0766$ with $z = -1.93$; P value = 0.0268; Do not reject H_0.

14. H_0: $p = 0.238$; H_1: $p > 0.238$; right-tailed; normal $z_0 = 2.33$; $\hat{p} = 0.2918$ with $z = 3.12$; P value = 0.0009; Reject H_0.

Section 9.6

2. H_0: $\mu_d = 0$; H_1: $\mu_d \neq 0$; two-tailed; $t_0 = \pm 3.707$; $\bar{d} = 0.37$; $s_d = 0.47$; sample t = 2.08; $0.05 < $ P value < 0.10; Fail to reject H_0.

4. H_0: $\mu_d = 0$; H_1: $\mu_d > 0$; right-tailed; $t_0 = 2.821$; $\bar{d} = 0.08$; $s_d = 1.701$; sample t = 0.1487; P value > 0.125; Fail to reject H_0.

6. H_0: $\mu_d = 0$; H_1: $\mu_d \neq 0$; two-tailed; $t_0 = \pm 3.106$; $\bar{d} = -0.84$; $s_d = 3.57$; sample t = -0.815; P value > 0.250; Fail to reject H_0.

8. H_0: $\mu_d = 0$; H_1: $\mu_d > 0$; right-tailed; $t_0 = 1.943$; $\bar{d} = 0.0$; $s_d = 8.76$; sample t = 0.000; P value > 0.125; Fail to reject H_0.

10. H_0: $\mu_d = 0$; H_1: $\mu_d > 0$; right-tailed; $t_0 = 1.895$; $\bar{d} = 1.25$; $s_d = 1.91$; sample t = 1.851; $0.050 < $ P value < 0.075; Fail to reject H_0.

12. H_0: $\mu_d = 0$; H_1: $\mu_d \neq 0$; two-tailed; $t_0 = \pm 2.571$; $\bar{d} = -3.33$; $s_d = 7.34$; sample t = -1.11; P value > 0.25; Fail to reject H_0.

14. H_0: $\mu_d = 0$; H_1: $\mu_d > 0$; right-tailed; $t_0 = 2.015$; $\bar{d} = 0.4$; $s_d = 0.447$; sample t = 2.192; $0.025 < $ P value < 0.05; Reject H_0.

16. H_0: $\mu_d = 0$; H_1: $\mu_d \neq 0$; two-tailed; $t_0 = \pm 2.306$; $\bar{d} = 0.1111$; $s_d = 5.2784$; sample t = 0.0631; P value > 0.25; Fail to reject H_0.

Section 9.7

2. H_0: $\mu_1 = \mu_2$; H_1: $\mu_1 > \mu_2$; $z_0 = 2.33$; $z = 3.50$ for $\bar{x}_1 - \bar{x}_2 = 13$; P value = 0.0002; Reject H_0

4. H_0: $\mu_1 = \mu_2$; H_1: $\mu_1 > \mu_2$; $z_0 = 2.33$; $z = 1.73$ for $\bar{x}_1 - \bar{x}_2 = 0.3$; P value = 0.0418; Fail to reject H_0.

6. H_0: $\mu_1 = \mu_2$; H_1: $\mu_1 \neq \mu_2$; $z_0 = \pm 2.58$; $z = -0.86$ for $\bar{x}_1 - \bar{x}_2 = -0.4$; P value = 0.3898; Fail to reject H_0

8. H_0: $\mu_1 = \mu_2$; H_1: $\mu_1 > \mu_2$; $z_0 = 2.33$; $z = 1.52$ for $\bar{x}_1 - \bar{x}_2 = 19.2$; P value = 0.0643; Fail to reject H_0

Section 9.7 continued

10. (a) Use calculator.
    (b) H_0: $\mu_1 = \mu_2$; H_1: $\mu_1 > \mu_2$; d.f. = 18; $t_0 = 1.734$; $s = 18.17$; $t = 2.102$ for $\bar{x}_1 - \bar{x}_2 =$ 17.17; $0.010 < $ P value < 0.025; Reject H_0

12. (a) Use calculator.
    (b) H_0: $\mu_1 = \mu_2$; H_1: $\mu_1 < \mu_2$; d.f. = 12; $t_0 = -1.782$; $s = 2.5397$; $t = -0.214$ for $\bar{x}_1 - \bar{x}_2 =$ -0.29; P value > 0.125; Fail to reject H_0

14. H_0: $\mu_1 = \mu_2$; H_1: $\mu_1 < \mu_2$; $t_0 = -2.492$; $s = 7.895$; $t = -1.914$ for $\bar{x}_1 - \bar{x}_2 = -6$; $0.025 < $ P value < 0.05; Fail to reject H_0.

16. (a) Use calculator.
    (b) H_0: $\mu_1 = \mu_2$; H_1: $\mu_1 < \mu_2$; $t_0 = -1.812$; $s = 2.8$; $t = -1.732$ for $\bar{x}_1 - \bar{x}_2 = -2.8$; $0.05 < $ P value < 0.075; Fail to reject H_0

18. H_0: $p_1 = p_2$; H_1: $p_1 < p_2$; $z_0 = -1.645$; $\hat{p} = 0.5278$; $z = -1.98$ for $\hat{p}_1 - \hat{p}_2 = -0.0891$; P value $= 0.0239$; Reject H_0.

20. H_0: $p_1 = p_2$; H_1 $p_1 > p_2$; $z_0 = 2.33$; $\hat{p} = 0.4900$; $z = 8.204$ for $\hat{p}_1 - \hat{p}_2 = 0.5800$; P value < 0.0001; reject H_0

22. H_0: $p_1 = p_2$; H_1: $p_1 \neq p_2$; $z_0 = \pm 1.96$; $\hat{p} = 0.0234$; $z = -0.18$ for $\hat{p}_1 - \hat{p}_2 =$ -0.0020; P value $= 0.8572$; Fail to reject H_0.

Chapter 9 Review Problems

2. Single proportion; H_0: $p = 0.35$; H_1: $p > 0.35$; $z_0 = 1.645$; $z = 2.48$ for $\hat{p} = 0.4815$; P value $= 0.0066$; Reject H_0

4. Difference of means; small independent samples; H_0: $\mu_1 = \mu_2$; H_1: $\mu_1 > \mu_2$; d.f. = 22; $t_0 = 2.508$; $s = 2.0506$; $t = 3.106$ for $\bar{x}_1 - \bar{x}_2 = 2.6$; P value < 0.005; Reject H_0

6. Single mean; large sample; H_0: $\mu = \$29,800$; H_1: $\mu < \$29,800$; $z_0 = -1.645$; $z = -8.00$ for $\bar{x} = 29,000$; P value < 0.0001; Reject H_0.

8. Difference of means, large samples: H_0: $\mu_1 = \mu_2$; H_1: $\mu_1 \neq \mu_2$; $z_0 = \pm 1.96$; $z = -3.48$ for $\bar{x}_1 - \bar{x}_2 = -9$; P value$= 0.0006$; Reject H_0.

10. Difference of proportions, $\hat{p}_1 = 0.1364$; $\hat{p}_2 = 0.1856$; H_0: $p_1 = p_2$; H_1: $p_1 < p_2$; $z_0 = -1.645$; $z = -0.91$; P value $= 0.1814$; Fail to reject H_0

12. Single Proportion; H_0: $p = 0.36$; H_1: $p < 0.36$; $z_0 = -1.645$; $z = -1.94$ for $\hat{p} = 33/120$; P value $= 0.0262$; Reject H_0.

14. Paired difference test; H_0: $\mu_d = 0$; H_1: $\mu_d > 0$; d.f. = 5; $t_0 = 3.365$; $t = 6.066$ for $\bar{d} = 9.833$; P value < 0.005; Reject H_0

16. (a) Use calculator.
    (b) Single mean; small sample; H_0: $\mu = 48$; H_1: $\mu < 48$; d.f. = 9; $t_0 = -1.833$; $t = -0.525$; P value > 0.125; Fail to reject H_0

18. (a) Do not reject H_0; P value > 0.01
    (b) Reject H_0; P value < 0.05

Chapter 10

Section 10.1

2. No liner correlation
4. Moderate Linear correlation
6. No linear correlation
8.

(a) List Price and Dealer's Invoice for Chevrolet Cavalier (thousands of dollars)

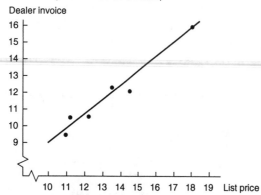

(b) Draw line you think best. (Specific equation for line in found in Section 10.2)
(c) Moderate

10.

(a) Group Health Insurance Plans: Average Number of Employees Versus Administrative Costs as a Percentage of Claims

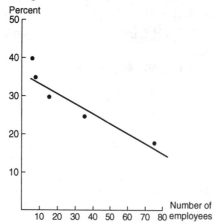

(b) Draw line you think best. (Specific equation for line in found in Section 10.2)
(c) Moderate

Section 10.1 continued
12.

(a, b) Magnitude (Richter Scale) and Depth (km) of Earthquakes

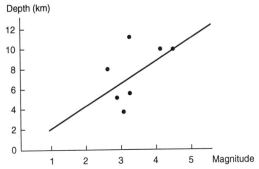

(c) Low

14.

(a, b) Body Weight and Metabolic Rate of Children

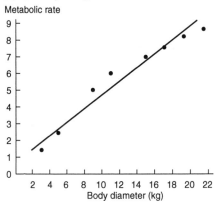

(c) High

Section 10.2
2.

(a, c) Age and Weight of Healthy Calves

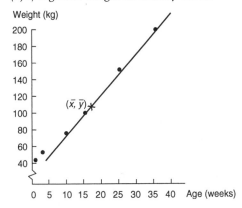

Section 10.2 continued

2. (b) $\bar{x} = 15.33$ weeks; $\bar{y} = 102.83$ kg; $b = 4.50899$; $y = 33.70 + 4.51x$
    (d) $S_e = 4.6706$ (e) $y = 87.6$ (f) 76.99 kg to 98.6 kg

4.

(a, c) Fouls and Basketball Losses

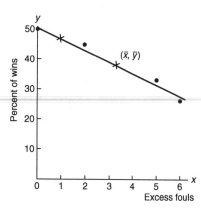

(b) $\bar{x} = 3.25$; $\bar{y} = 38.5$; $b = -3.934066$; $y = -3.934x + 51.29$
(d) $S_e = 2.109632$ (e) 35.55   (f) 31.0 to 40.04

6.

(a, c) Percentage of 16- to 19-Year-Olds Not in
School and Per Capita Income (thousands of dollars)

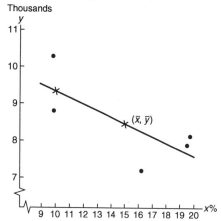

(b) $\bar{x} = 15.02$; $\bar{y} = 8.46$; $b = -0.175838$; $y = -0.176x + 11.10$
(d) $S_e = 0.919816$ (e) 8.11   (f) 6.65 to 9.57

Section 10.2 continued

8.

(a, c) List Price and Dealer Invoice for GMC Sonoma
(thousands of dollars)

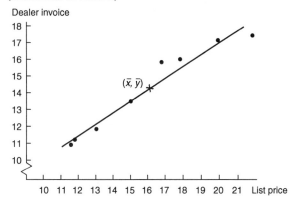

(b) $\bar{x} = 15.96$; $\bar{y} = 14.24$; $b = 0.7107$; $y = 2.89 + 0.71x$
(d) $S_e = 0.5919$　　　(e) 13.2　　　(f) 12.4 to 14.0

10.

(a, c) Number of Research Programs and Mean
Number of Patents per Program

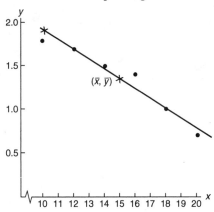

(b) $\bar{x} = 15.0$; $\bar{y} = 1.35$; $b = -0.1100$; $y = -0.11x + 3.0$
(d) $S_e = 0.109545$　　　(e) 1.35　　　(f) 1.14 to 1.56

Section 10.2 continued

12.

(a, c) Ages of Children and Their Responses to Questions

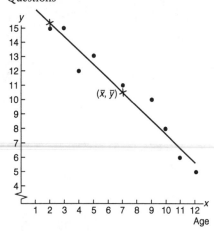

(b) $\bar{x} = 7.0$; $\bar{y} = 10.56$; $b = -0.962963$; $y = -0.963x + 17.30$

(d) $S_e = 0.931518$ (e) 8.15   (f) 4.6 to 11.7

14.

(a, c) Body Weight (kg) and Metabolic Rate (100 Kcal/24 h)

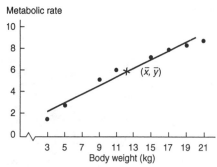

(b) $\bar{x} = 12.5$; $\bar{y} = 5.8875$; $b = 0.40248$; $y = 0.8565 + 0.4025x$

(d) $S_e = 0.5175$ (e) 7.3 Kcal/24 hr   (f) 6.6 to 8.0

16. (a) Result checks.   (b) Result checks   (c) Yes

     (d) The equation $x = 0.9337y - 0.1335$ does not match part b.

     (e) In general, switching x and y values produces a *different* least-squares equation. It is important that when you perform a linear regression you know which variable is the explanatory variable and which is the response variable.

Section 10.2 continued

18.

(a) Residuals: 2.9; 2.1; −0.1; −2.1; −0.5; −2.3; −1.9; 1.9.
Residual Plot

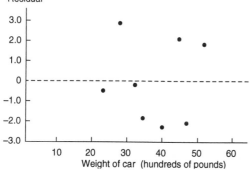

(b) Residuals seem to be scattered randomly around the horizontal line at 0. No outliers.

Section 10.3

2. (a) No, we hope not!   (b) Increase in buying power due to increase in salaries
4. (a) No
   (b) Increase in population could account for increases in both consumption of soda pop and in number of traffic accidents.

6.

(a) Leisure Airfare and Business Airfare (dollars)

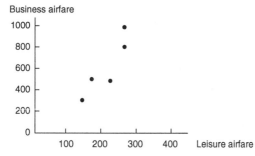

(b) Closest to 1   (c) $r = 0.889$; $r^2 = 0.790$; 79% explained; 21% unexplained

Section 10.3 continued

8.

(a) Per Capita Income and Death Rates in Small Cities in Oregon

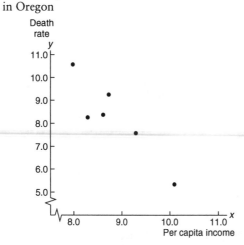

(b) -1   (c) r = -0.919; r^2 = 0.845; 84.5% explained; 15.5% unexplained

10.

(a) Percentage of 16- to 19-Year-Olds Not in School and Death Rate per 1000 Residents

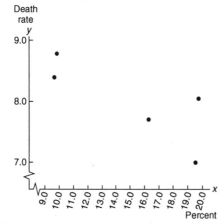

(b) -1   (c) r = -0.778; r^2 = 0.605; 60.5% explained; 39.5% unexplained

Section 10.3 continued

12.

(a) Driver's Age and Fatal Accident Rate Due to Not Yielding

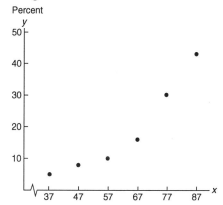

(b) 1   (c) $r = 0.943$; $r^2 = 0.889$; 88.9% explained; 11.1% unexplained

14.

(a) Lowest Barometric Pressure and Maximum Wind Speed for Tropical Cyclones

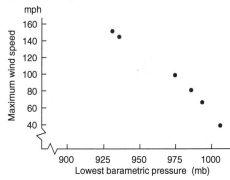

(b) Closest to -1   (c) $r = -0.9897$; $r^2 = 0.9796$; 98% explained; 2% unexplained

16. (a) $SS_{xy} = SS_{yx}$ (b) Same   (c) Same

(d) $r = 0.6186$ is both case; least-squares equations are not necessarily the same.

Section 10.4

2. H_0: $\rho = 0$; H_1: $\rho > 0$; $r_0 = 0.44$; $r = 0.646$; reject H_0

4. H_0: $\rho = 0$; H_1: $\rho > 0$; $r_0 = 0.54$; $r = 0.737$; reject H_0

6. H_0: $\rho = 0$; H_1: $\rho > 0$; $r_0 = 0.93$; $r = 0.999$; reject H_0

8. H_0: $\rho = 0$; H_1: $\rho \neq 0$; $r_0 = \pm0.87$; $r = 0412$; fail to reject H_0

10. H_0: $\rho = 0$; H_1: $\rho < 0$; $r_0 = -0.75$; $r = -0.60$; fail to reject H_0

12. H_0: $\rho = 0$; H_1: $\rho > 0$; $r_0 = 0.48$; $r = 0.71$; reject H_0

Section 10.5

2. (a) Response variable is x_3; explanatory variables are x_1, x_4, x_7
    (b) Constant term is -16.5; 4.0 with x_1; 9.2 with x_4; -1.1 with x_7
    (c) 12.1   (d) 9.2 units; 27.6 units; -18.4 units
    (e) 7.75 to 10.85
    (f) H_0: $\beta_4 = 0$; H_1: $\beta_4 \neq 0$; $t_0 = \pm 3.106$; $t = 9.989$; reject H_0

4. See computer printout

6. See computer printout

Chapter 10 Review Problems

2.

(a, c) Annual Salary (thousands) and Number
of Job Changes

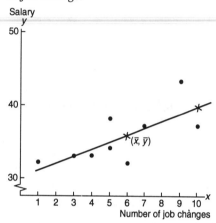

   (b) $\bar{x} = 6.0$; $\bar{y} = 35.9$; $b = 0.939024$; $y = 0.939x + 30.266$
   (d) 32.14   (e) $S_e = 2.564058$ (f) 26.72 to 37.57
   (g) Positive   (h) $r = 0.761$; $r^2 = 0.579$
   (i) H_0: $\rho = 0$; H_1: $\rho > 0$; $r_0 = 0.54$; reject H_0

4.

(a, c) Number of Insurance Sales and
Number of Visits

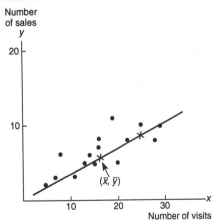

   (b) $\bar{x} = 16.53$; $\bar{y} = 6.47$; $b = 0.292748$; $y = 0.293x + 1.626$
   (d) $S_e = 1.730940$ (e) 6.896   (f) 3.73 to 10.07

Chapter 10 Review Problems continued

4. (h) H_0: $\rho = 0$; H_1: $\rho > 0$; $r_0 = 0.59$; reject H_0

6.

(a) Retail Price and "Fair" Price for
Auto Equipment (dollars)

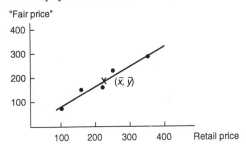

(b) Moderate   (c) $\bar{x} = 216$; $\bar{y} = 176.2$; $b = 0.81116$; $y = 0.99 + 0.81x$
(d) $r = 0.974$; $r^2 = 0.949$; 95% explained; 5% unexplained
(e) \$142.74   (f) 87.38 to 198.10
(g) H_0: $\rho = 0$; H_1: $\rho > 0$; $r_0 = 0.81$; The sample statistic r lies in the critical region and is significant; reject H_0.

Chapter 11

Section 11.1

2. H_0: Meyers-Briggs preference and profession are independent; H_1: Meyers-Briggs preference and profession are not independent; $\chi^2 = 174.6$; d.f. = 2; $\chi^2_{0.01} = 9.21$; Reject H_0

4. H_0: Ceremonial ranking and pottery type are independent; H_1: Ceremonial ranking and pottery type are not independent; $\chi^2 = 19.6079$; d.f. = 2; $\chi^2_{0.05} = 5.99$; Reject H_0

6. H_0: Type and career choice are independent; H_1: Type and career choice are not independent; $\chi^2 = 15.6017$; d.f. = 6; $\chi^2_{0.01} = 16.81$; Do not reject H_0

8. H_0: Contribution and ethnic group are independent; H_1: Contribution and ethnic group are not independent; $\chi^2 = 190.44$; d.f. = 12; $\chi^2_{0.01} = 26.22$; Reject H_0

10. H_0: party affiliation and dollars spent are independent; H_1: Party affiliation and dollars spent are not independent; $\chi^2 = 2.17$; d.f = 2; $\chi^2 = 9.21$; Do not reject H_0

Section 11.2

2. H_0: the distributions are the same; H_1: the distributions are different; $\chi^2 = 13.017$; d.f. = 4; $\chi^2_{0.05} = 9.49$; Reject H_0

4. H_0: the distributions are the same; H_1: the distributions are different; $\chi^2 = 1.084$; d.f. = 4; $\chi^2_{0.05} = 9.49$; Do not reject H_0

6. (a) Essay   (b) H_0: the distribution is normal; H_1: the distribution is not normal; $\chi^2 = 0.4501$; d.f. = 5; $\chi^2_{0.01} = 15.09$; Do not reject H_0

Section 11.2 continued

8. H_0: the distributions are the same; H_1: the distributions are different; $\chi^2 = 91.51$; d.f. = 4; $\chi^2_{0.05} = 9.49$; Reject H_0.

10. H_0: the distributions are the same; H_1: the distributions are different; $\chi^2 = 15.65$; d.f. = 5; $\chi^2_{0.01} = 15.09$; reject H_0

Section 11.3

2. H_0: $\sigma^2 = 5.1$; H_1: $\sigma^2 < 5.1$; d.f. = 40; $\chi^2_{0.95} = 26.51$; $\chi^2 = 25.88$; Reject H_0. Interval from 2.37 to 4.98.

4. H_0: $\sigma^2 = 47.1$; H_1: $\sigma^2 > 47.1$; d.f. = 14; $\chi^2_{0.05} = 23.68$; $\chi^2 = 24.73$; Reject H_0. Interval from 44.59 to 206.89

6. H_0: $\sigma^2 = 225$; H_1: $\sigma^2 > 225$; d.f. = 9; $\chi^2_{0.01} = 21.67$; $\chi^2 = 23.04$; Reject H_0. Interval from 16.5 to 43.8

8. (a) H_0: $\sigma^2 = 5625$; H_1: $\sigma^2 \neq 5625$; critical values are 44.18 and 9.26; $\chi^2 = 21.20$; Do not reject H_0 (b) $2698.68 < \sigma^2 < 12876.03$ (c) $51.95 < \sigma < 113.47$

Section 11.4

2. Population one is annual production from the second plot. $d.f._N = 7$; $d.f._D = 10$; H_0: $\sigma_1^2 = \sigma_2^2$; H_1: $\sigma_1^2 \neq \sigma_2^2$; sample F = 3.39; $F_0 = 3.95$; Do not reject H_0.

4. Population one refers to data of South Korean companies. $d.f._N = 12$; $d.f._D = 8$ H_0: $\sigma_1^2 = \sigma_2^2$; H_1: $\sigma_1^2 > \sigma_2^2$; sample F = 3.60; $F_0 = 3.28$; Reject H_0.

6. Population one refers to data from intermediate term bonds. $d.f._N = 15$; $d.f._D = 12$ H_0: $\sigma_1^2 = \sigma_2^2$; H_1: $\sigma_1^2 \neq \sigma_2^2$; sample F = 5.30; $F_0 = 3.18$; Reject H_0.

8. Population one refers to data from the old thermostat. $d.f._N = 15$; $d.f._D = 20$ H_0: $\sigma_1^2 = \sigma_2^2$; H_1: $\sigma_1^2 > \sigma_2^2$; sample F = 2.51; $F_0 = 2.33$; Reject H_0.

Section 11.5

2. (a) H_0: $\mu_1 = \mu_2 = \mu_3 = \mu_4$; H_1: not all the means are equal
   (b-h)

Source of Variation	Sum of Squares	Degrees of Freedom	Mean Square	F Ratio	F Critical Value	Test Decision
Between groups	421.033	3	140.344	1.573	3.20	Do not reject H_0
Within groups	1516.967	17	89.233			
Total	1938.000	20				

Section 11.5 continued

4. (a) H_0: $\mu_1 = \mu_2 = \mu_3$; H_1: not all the means are equal
   (b-h)

Source of Variation	Sum of Squares	Degrees of Freedom	Mean Square	F Ratio	F Critical Value	Test Decision
Between groups	215.680	2	107.840	0.816	6.36	Do not reject H_0
Within groups	1981.725	15	132.115			
Total	2197.405	17				

6. (a) H_0: $\mu_1 = \mu_2 = \mu_3$; H_1: not all the means are equal
   (b-h)

Source of Variation	Sum of Squares	Degrees of Freedom	Mean Square	F Ratio	F Critical Value	Test Decision
Between groups	2.442	2	1.2208	2.95	3.55	Do not reject H_0
Within groups	7.448	18	0.4138			
Total	9.890	20				

8. (a) H_0: $\mu_1 = \mu_2 = \mu_3 = \mu_4$; Not all the means are equal.
   (b-h)

Source of Variation	Sum of Squares	Degrees of Freedom	Mean Square	F Ratio	F Critical Value	Test Decision
Between groups	18.965	3	6.322	14.910	3.41	Reject H_0
Within groups	5.517	13	0.424			
Total	24.482	16				

Section 11.6

2. 2 factors: rank with 4 levels and institution type with 2 levels; data table has 8 cells.
4. (a) 2 factors: education with 4 levels and media type with 5 levels
   (b) For education, H_0: No difference in population mean index according to education level; H_1: At least two education levels have different mean indices; $F_{education} = 2.963$; $F_{0.05} = 3.49$; Do not reject H_0 for education level.
   (c) For media, H_0: No difference in population mean index by media type; H_1: At least two types of media have different population mean indices; $F_{media} = 0.0093$; $F_{0.05} = 3.26$; Do not reject H_0.

Section 11.6 continued

6. (a) 2 factors: class with 4 levels and gender with 2 levels
   (b) H_0: No interaction between factors; H_1: Some interaction between factors; $F_{interaction}$ = 0.4072; $F_{0.05}$ = 3.01; Do not reject H_0.
   (c) H_0: No difference in population mean GPA based on class; H_1: At least two classes have different population mean GPA's. F_{class} = 3.2489; $F_{0.05}$ = 3.01; Reject H_0.
   (d) H_0: No difference in population mean GPA based on gender; H_1: Some difference in population mean GPA based on gender. F_{gender} = 1.3575; $F_{0.05}$ = 4.26; Do not reject H_0.

Chapter 11 Review Problems

2. H_0: Time to do a test and test score are independent. H_1: Time to do a test and test score are not independent. χ^2 = 3.92; $\chi^2_{0.01}$ = 11.34. Do not reject H_0; Time to do a test and test results are independent.

4. H_0: $\mu_1 = \mu_2 = \mu_3$; H_1: Not all the means are equal.

Source of Variation	Sum of Squares	Degrees of Freedom	Mean Square	F Ratio	F Critical Value	Test Decision
Between groups	1.002	2	0.501	0.443	8.02	Fail to reject H_0
Within groups	10.165	9	1.129			
Total	11.167	11				

6. H_0: $\sigma^2 = 0.0625$; H_1: $\sigma^2 > 0.0625$; critical value $\chi^2_{0.05}$ = 19.68. Since the observed value χ^2 = 25.41 fals in the critical region, we reject H_0; the machine needs to be adjusted.

8. H_0: $\sigma_1^2 = \sigma_2^2$; H_1: $\sigma_1^2 \neq \sigma_2^2$; sample F = 1.84; F_0 = 2.30; Do not reject H_0.

10. (a) 2 factors: day with 2 levels and section with 3 levels
    (b) H_0: No interaction between day and section; H_1: Some interaction between day and section; $F_{interaction}$ = 1.027; $F_{0.01}$ = 5.39; Do not reject H_0.
    (c) H_0: No difference in population mean number of responses according to day; H_1: At least two population means are different among the days; F_{day} = 55.35; $F_{0.01}$ = 7.56; Reject H_0.
    (d) H_0: No difference in population mean number of responses according to section; H_1: At least two population means are different among the sections; $F_{section}$ = 42.53; $F_{0.01}$ = 5.39; Reject H_0.

Chapter 12

Section 12.1

2. H_0: $\mu_1 = \mu_2$; H_1: $\mu_1 > \mu_2$; $z_0 = 1.645$; $z = 0.83$ for $r = 8/13$; Fail to reject H_0.
4. H_0: $\mu_1 = \mu_2$; H_1: $\mu_1 \neq \mu_2$; $z_0 = \pm 2.58$; $z = 0.53$ for $r = 8/14$; Fail to reject H_0.
6. H_0: $\mu_1 = \mu_2$; H_1: $\mu_1 < \mu_2$; $z_0 = -1.645$; $z = -0.69$ for $r = 8/19$; Fail to reject H_0.
8. H_0: $\mu_1 = \mu_2$; H_1: $\mu_1 < \mu_2$; $z_0 = -1.645$; $z = -2.32$ for $r = 3/15$; Reject H_0.
10. H_0: $\mu_1 = \mu_2$; H_1: $\mu_1 \neq \mu_2$; $z_0 = \pm 1.96$; $z = 1.39$ for $r = 9/13$; Fail to reject H_0.

Section 12.2

2. H_0: $\mu_1 = \mu_2$; H_1: $\mu_1 \neq \mu_2$; $z_0 = \pm 1.96$; $\mu_R = 115$; $\sigma_R = 15.17$; $R = 117$; $z = 0.13$;
   Fail to reject H_0.
4. H_0: $\mu_1 = \mu_2$; H_1: $\mu_1 \neq \mu_2$; $z_0 = \pm 1.96$; $\mu_R = 105$; $\sigma_R = 13.23$; $R = 92$; $z = -0.98$;
   Fail to reject H_0.
6. H_0: $\mu_1 = \mu_2$; H_1: $\mu_1 \neq \mu_2$; $z_0 = \pm 2.58$; $\mu_R = 110$; $\sigma_R = 14.20$; $R = 97.5$; $z = -0.88$
   Fail to reject H_0.
8. H_0: $\mu_1 = \mu_2$; H_1: $\mu_1 \neq \mu_2$; $z_0 = \pm 2.58$; $\mu_R = 85.5$; $\sigma_R = 11.32$; $R = 99$; $z = 1.19$;
   Fail to reject H_0.
10. H_0: $\mu_1 = \mu_2$; H_1: $\mu_1 \neq \mu_2$; $z_0 = \pm 2.58$; $\mu_R = 68$; $\sigma_R = 9.52$; $R = 75.5$; $z = 0.79$;
    Fail to reject H_0.

Section 12.3

2. H_0: $\rho_S = 0$; H_1: $\rho_S \neq 0$; critical values 0.680 and -0.680. Since the observed value $r_S = 0.349$ falls outside the critical region, we do not reject H_0; There is no monotone relation between costs and earnings.
4. H_0: $\rho_S = 0$; H_1: $\rho_S > 0$; critical value is 0.564. Since the observed value $r_S = 0.488$ is outside the critical region, we do not reject H_0; There is no monotone relation between rank of finish and rank of score.
6. H_0: $\rho_S = 0$; H_1: $\rho_S \neq 0$; critical values 0.715 and -0.715. Since the observed value $r_S = 0.452$ falls outside the critical region, we do not reject H_0; There is no monotone relation between quality rank and price rank.
8. H_0: $\rho_S = 0$; H_1: $\rho_S > 0$; critical value is 0.829. Since the observed value $r_S = 0.257$ falls outside the critical region, we fail to reject H_0; There is no monotone relation between the opinions of the managers.

Chapter 12 Review Problems

2. H_0: $\mu_1 = \mu_2$; H_1: $\mu_1 > \mu_2$; $z_0 = 1.645$; $z = 2.17$ for $r = 0.79$; Reject H_0
4. H_0: $\mu_1 = \mu_2$; H_1: $\mu_1 \neq \mu_2$; $z_0 = \pm 2.58$; $\mu_R = 90$; $\sigma_R = 12.25$; $R = 116$; $z = 2.12$;
   Reject H_0.
6. H_0: $\rho_S = 0$; H_1: $\rho_S \neq 0$; critical values 0.900 and -0.900. Since the observed value $r_S = -0.700$ does not fall in the critical region, we do not reject H_0; There is no monotone relation between opinions of the chefs.

Part IV

Transparency Masters

RANDOM NUMBERS

1st Line
92630   78240   19267   95457   53497
23894   37708   79862   76471   66418

2nd Line
79445   78735   71549   44843   26104
67318   00701   34986   66751   99723

3rd Line
59654   71966   27386   50004   05358
94031   29281   18544   52429   06080

4th Line
31524   49587   76612   39789   13537
48086   59483   60680   84675   53014

5th Line
06348   76938   90379   51392   55887
71015   09209   79157   24440   30244

Figure 2-14

Minitab Generated

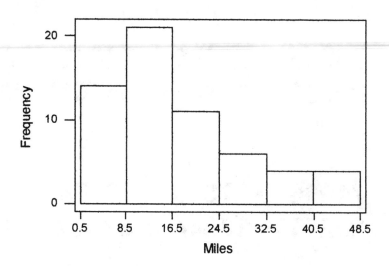

Figure 2-26

Minitab Generated

Ogive for Daily High Temperature

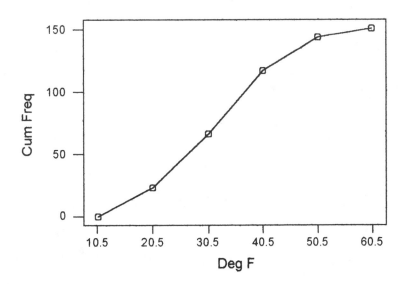

Figure 2-32

Weight of Carry-On Luggage

3| 2 represents 32 LB

Stem	Leaf
0	3 0
1	2 7 8 8 9 2 8
2	7 7 2 9 1 6 1 8 9 1 6
3	0 5 8 6 5 8 2 3 2 1 2 3 1 2
4	2 7 1 5 3
5	1

Figure 3-12

TI-83 Generated

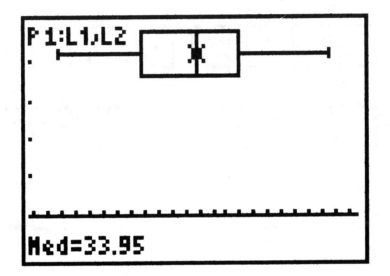

Sample Space for Throwing Two Dice

Sample Space for Drawing One Card
From a Standard Bridge Deck

Red Cards		Black Cards	
Hearts	Diamonds	Clubs	Spades
2	2	2	2
3	3	3	3
4	4	4	4
5	5	5	5
6	6	6	6
7	7	7	7
8	8	8	8
9	9	9	9
10	10	10	10
Jack	Jack	Jack	Jack
Queen	Queen	Queen	Queen
King	King	King	King
Ace	Ace	Ace	Ace

Figure 6-5
Areas Under a Normal Curve

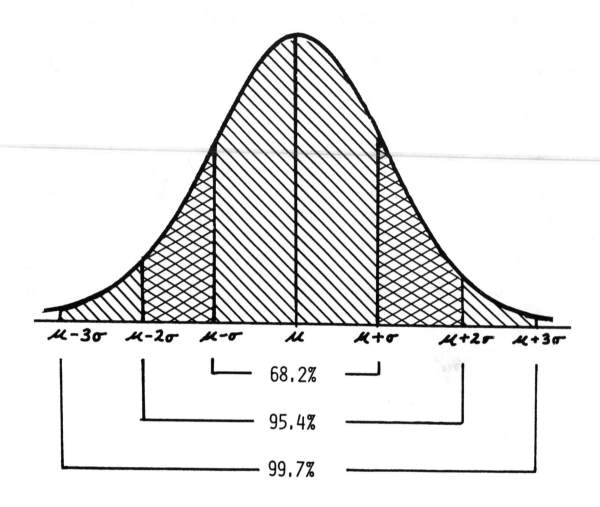

Figure 6-9
Control Chart for Number of Rooms Not Made Up
ComputerStat Generated

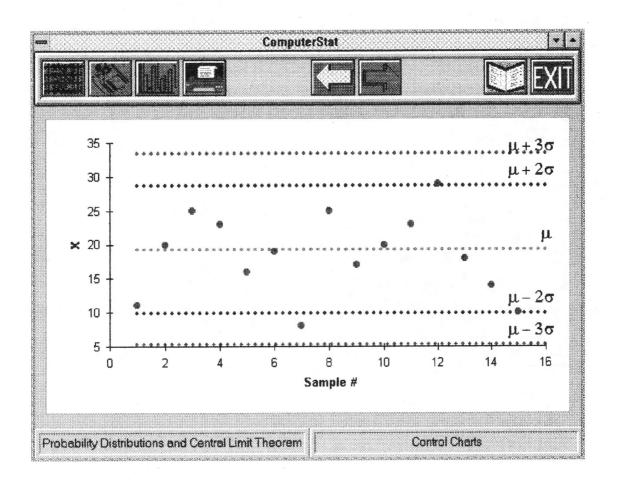

Section 9.4 Problem #7
ComputerStat Display

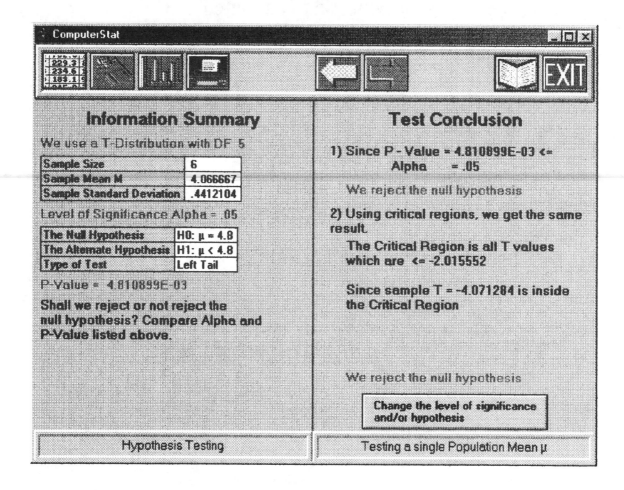

Information Summary

We use a T-Distribution with DF 5

Sample Size	6
Sample Mean M	4.066667
Sample Standard Deviation	.4412104

Level of Significance Alpha = .05

The Null Hypothesis	H0: $\mu = 4.8$
The Alternate Hypothesis	H1: $\mu < 4.8$
Type of Test	Left Tail

P-Value = 4.810899E-03

Shall we reject or not reject the null hypothesis? Compare Alpha and P-Value listed above.

Hypothesis Testing

Test Conclusion

1) Since P - Value = 4.810899E-03 <= Alpha = .05

We reject the null hypothesis

2) Using critical regions, we get the same result.

The Critical Region is all T values which are <= -2.015552

Since sample T = -4.071284 is inside the Critical Region

We reject the null hypothesis

Change the level of significance and/or hypothesis

Testing a single Population Mean μ

Least Squares Line and Scatter Diagram
Figure 10-8
ComputerStat Generated

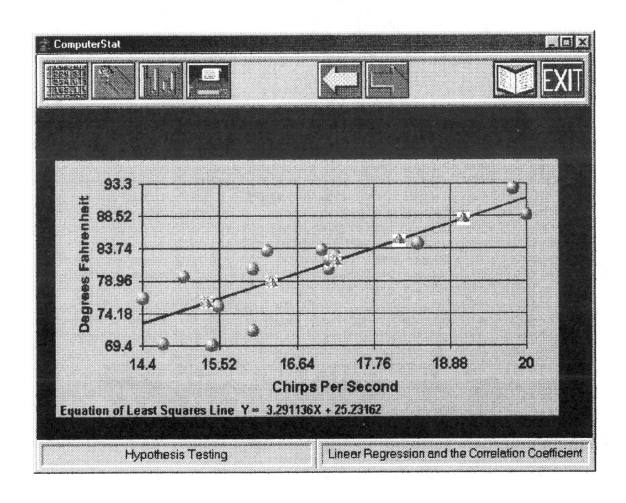

Figure 10-12
95% Prediction Band
Minitab Generated

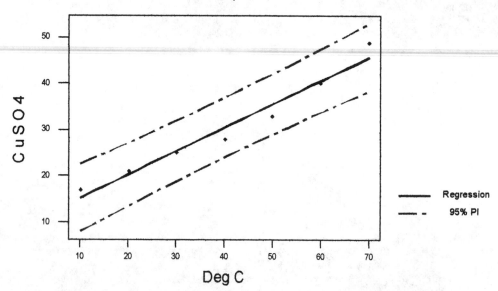

Chi-Square Test
Table 11-2 Keyboard Versus Time to Learn to Type
Minitab Generated

Expected counts are printed below observed counts

	21-41 hr	41-60 hr	61-80 hr	Total
1	25	30	25	80
	24.00	40.00	16.00	
2	30	71	19	120
	36.00	60.00	24.00	
3	35	49	16	100
	30.00	50.00	20.00	
Total	90	150	60	300

Chi-Sq = 0.042 + 2.500 + 5.062 +
         1.000 + 2.017 + 1.042 +
         0.833 + 0.020 + 0.800 = 13.316

DF = 4, P-Value = 0.010

Part V

Formula Card Masters

Masters for Statistical Tables

FREQUENTLY USED FORMULAS

n = sample size N = population size f = frequency

Chapter 2

Class Width = $\dfrac{\text{high} - \text{low}}{\text{number classes}}$ (increase to next integer)

Class Midpoint = $\dfrac{\text{upper limit} + \text{lower limit}}{2}$

Lower boundary = lower boundary of previous class + class width

Chapter 3

Sample mean $\bar{x} = \dfrac{\Sigma x}{n}$

Population mean $\mu = \dfrac{\Sigma x}{N}$

Weighted average = $\dfrac{\Sigma xw}{\Sigma w}$

Range = largest data value − smallest data value

Sample standard deviation $s = \sqrt{\dfrac{\Sigma(x - \bar{x})^2}{n - 1}}$

Computation formula $s = \sqrt{\dfrac{SS_x}{n - 1}}$ where

$SS_x = \Sigma x^2 - \dfrac{(\Sigma x)^2}{n}$

Population standard deviation $\sigma = \sqrt{\dfrac{\Sigma(x - \mu)^2}{N}}$

Sample variance s^2

Population variance σ^2

Sample Coefficient of Variation $CV = \dfrac{s}{\bar{x}} \cdot 100$

Sample mean for grouped data $\bar{x} = \dfrac{\Sigma xf}{n}$

Sample standard deviation for grouped data

$s = \sqrt{\dfrac{\Sigma(x - \bar{x})^2 f}{n - 1}} = \sqrt{\dfrac{\Sigma x^2 f - (\Sigma xf)^2/n}{n - 1}}$

Chapter 4

Probability of the complement of event A
$P(not\ A) = 1 - P(A)$

Multiplication rule for independent events
$P(A\ and\ B) = P(A) \cdot P(B)$

General multiplication rules
$P(A\ and\ B) = P(A) \cdot P(B,\ given\ A)$
$P(A\ and\ B) = P(B) \cdot P(A,\ given\ B)$

Addition rule for mutually exclusive events
$P(A\ or\ B) = P(A) + P(B)$

General addition rule
$P(A\ or\ B) = P(A) + P(B) - P(A\ and\ B)$

Permutation rule $P_{n,r} = \dfrac{n!}{(n - r)!}$

Combination rule $C_{n,r} = \dfrac{n!}{r!(n - r)!}$

Chapter 5

Mean of a discrete probability distribution $\mu = \Sigma x P(x)$

Standard deviation of a discrete probability distribution
$\sigma = \sqrt{\Sigma(x - \mu)^2 P(x)}$

For Binomial Distributions

r = number of successes; p = probability of success;
$q = 1 - p$

Binomial probability distribution $P(r) = \dfrac{n!}{r!(n - r)!} p^r q^{n-r}$

Mean $\mu = np$

Standard deviation $\sigma = \sqrt{npq}$

Geometric Probability Distribution

n = number of trial on which first success occurs
$P(n) = p(1 - p)^{n-1}$

Poisson Probability Distribution

λ = mean number of successes over given interval
$P(\lambda) = \dfrac{e^{-\lambda}\lambda^r}{r!}$

Chapter 6

Raw score $x = z\sigma + \mu$

Standard score $z = \dfrac{x - \mu}{\sigma}$

Chapter 7

Mean of $\bar{x}$ distribution $\mu_{\bar{x}} = \mu$

Standard deviation of $\bar{x}$ distribution $\sigma_{\bar{x}} = \dfrac{\sigma}{\sqrt{n}}$

Standard score for $\bar{x}$ $z = \dfrac{\bar{x} - \mu}{\sigma/\sqrt{n}}$

Chapter 8

Confidence Interval

for $\mu(n \geq 30)$

$$\bar{x} - z_c \frac{\sigma}{\sqrt{n}} < \mu < \bar{x} + z_c \frac{\sigma}{\sqrt{n}}$$

for $\mu(n < 30)$

$$d.f. = n - 1$$

$$\bar{x} - t_c \frac{s}{\sqrt{n}} < \mu < \bar{x} + t_c \frac{s}{\sqrt{n}}$$

for $p(np > 5 \text{ and } nq > 5)$

$$\hat{p} - z_c \sqrt{\frac{\hat{p}(1 - \hat{p})}{n}} < p < \hat{p} + z_c \sqrt{\frac{\hat{p}(1 - \hat{p})}{n}}$$

where $\hat{p} = r/n$

for difference of means ($n_1 \geq 30$ and $n_2 \geq 30$)

$$(\bar{x}_1 - \bar{x}_2) - z_c \sqrt{\frac{\sigma_1^2}{n_1} + \frac{\sigma_2^2}{n_2}} < \mu_1 - \mu_2 < (\bar{x}_1 - \bar{x}_2)$$

$$+ z_c \sqrt{\frac{\sigma_1^2}{n_1} + \frac{\sigma_2^2}{n_2}}$$

for difference of means ($n_1 < 30$ and/or $n_2 < 30$ and $\sigma_1 \approx \sigma_2$)

$$d.f. = n_1 + n_2 - 2$$

$$(\bar{x} - \bar{x}_2) - t_c s \sqrt{\frac{1}{n_1} + \frac{1}{n_2}} < \mu_1 - \mu_2 < (\bar{x}_1 - \bar{x}_2)$$

$$+ t_c s \sqrt{\frac{1}{n_1} + \frac{1}{n_2}}$$

where $s = \sqrt{\dfrac{(n_1 - 1)s_1^2 + (n_2 - 1)s_2^2}{n_1 + n_2 - 2}}$

for difference of proportions

where $\hat{p}_1 = r_1/n_1$; $\hat{p}_2 = r_2/n_2$; $\hat{q}_1 = 1 - \hat{p}_1$; $\hat{q}_2 = 1 - \hat{p}_2$

$$(\hat{p}_1 - \hat{p}_2) - z_c \sqrt{\frac{\hat{p}_1\hat{q}_1}{n_1} + \frac{\hat{p}_2\hat{q}_2}{n_2}} < p_1 - p_2 < (\hat{p}_1 - \hat{p}_2)$$

$$+ z_c \sqrt{\frac{\hat{p}_1\hat{q}_1}{n_1} + \frac{\hat{p}_2\hat{q}_2}{n_2}}$$

Sample Size for Estimating

means $n = \left(\dfrac{z_c \sigma}{E}\right)^2$

proportions

$n = p(1 - p)\left(\dfrac{z_c}{E}\right)^2$ with preliminary estimate for p

$n = \dfrac{1}{4}\left(\dfrac{z_c}{E}\right)^2$ without preliminary estimate for p

Chapter 9

Sample Test Statistics for Tests of Hypotheses

for $\mu(n \geq 30)$ $z = \dfrac{\bar{x} - \mu}{\sigma\sqrt{n}}$

for $\mu(n < 30)$; $d.f. = n - 1$ $t = \dfrac{\bar{x} - \mu}{s/\sqrt{n}}$

for p $z = \dfrac{\hat{p} - p}{\sqrt{pq/n}}$ where $q = 1 - p$

for paired differences d $t = \dfrac{\bar{d} - \mu_d}{s_d/\sqrt{n}}$ with $d.f. = n - 1$

difference of means large sample

$$z = \frac{(\bar{x}_1 - \bar{x}_2) - (\mu_1 - \mu_2)}{\sqrt{\dfrac{\sigma_1^2}{n_1} + \dfrac{\sigma_2^2}{n_2}}}$$

difference of means small sample with $\sigma_1 \approx \sigma_2$;

$$d.f. = n_1 + n_2 - 2$$

$$t = \frac{(\bar{x}_1 - \bar{x}_2) - (\mu_1 - \mu_2)}{s\sqrt{\dfrac{1}{n_1} + \dfrac{1}{n_2}}}$$

where $s = \sqrt{\dfrac{(n_1 - 1)s_1^2 + (n_2 - 1)s_2^2}{n_1 + n_2 - 2}}$

difference of proportions

$$z = \frac{\hat{p}_1 - \hat{p}_2}{\sqrt{\dfrac{\bar{p}\bar{q}}{n_1} + \dfrac{\bar{p}\bar{q}}{n_2}}}$$ where $\bar{p} = \dfrac{r_1 + r_2}{n_1 + n_2}$; $\bar{q} = 1 - \bar{p}$;

$$\hat{p}_1 = r_1/n_1; \hat{p}_2 = r_2/n_2$$

Chapter 10

Regression and Correlation

In all these formulas $SS_x = \Sigma x^2 - \dfrac{(\Sigma x)^2}{n}$,

$SS_y = \Sigma y^2 - \dfrac{(\Sigma y)^2}{n}$, $SS_{xy} = \Sigma xy - \dfrac{(\Sigma x)(\Sigma y)}{n}$

Least squares line $y = a + bx$ where $b = \dfrac{SS_{xy}}{SS_x}$ and

$a = \bar{y} - b\bar{x}$

Standard error of estimate $S_e = \sqrt{\dfrac{SS_y - bSS_{xy}}{n-2}}$

where $b = \dfrac{SS_{xy}}{SS_x}$

Pearson product moment correlation coefficient

$r = \dfrac{SS_{xy}}{\sqrt{SS_x SS_y}}$

Coefficient of determination $= r^2$

Confidence interval for y

$y_p - E < y < y_p + E$ where y_p is the predicted y value for x

$E = t_c S_e \sqrt{1 + \dfrac{1}{n} + \dfrac{(x - \bar{x})^2}{SS_x}}$ with $d.f. = n - 2$

Chapter 11

$x_2 = \Sigma \dfrac{(O - E)^2}{E}$ where $E = \dfrac{(\text{row total})(\text{column total})}{\text{sample size}}$

Tests of Independence $d.f. = (R - 1)(C - 1)$

Goodness of fit $d.f. = (\text{number of entries}) - 1$

Confidence Interval for σ^2; $d.f. = n - 1$

$\dfrac{(n - 1)s^2}{\chi_U^2} < \sigma^2 < \dfrac{(n - 1)s^2}{\chi_L^2}$

Sample test statistic for H_0: $\sigma^2 = k$; $d.f. = n - 1$

$\chi^2 = \dfrac{(n - 1)s^2}{\sigma^2}$

Testing Two Variances

Sample test statistic $F = \dfrac{s_1^2}{s_2^2}$

where $s_1^2 \geq s_2^2$

$d.f._N = n_1 - 1$; $d.f._D = n_2 - 1$

ANOVA

k = number of groups; N = total sample size

$SS_{TOT} = \Sigma x_{TOT}^2 - \dfrac{(\Sigma x_{TOT})^2}{N}$

$SS_{BET} = \sum_{all\ groups} \left(\dfrac{(\Sigma x_i)^2}{n_i} \right) - \dfrac{(\Sigma x_{TOT})^2}{N}$

$SS_W = \sum_{all\ groups} \left(\Sigma x_i^2 - \dfrac{(\Sigma x_i)^2}{n_i} \right)$

$SS_{TOT} = SS_{BET} + SS_W$

$MS_{BET} = \dfrac{SS_{BET}}{d.f._{BET}}$ where $d.f._{BET} = k - 1$

$MS_W = \dfrac{SS_W}{d.f._W}$ where $d.f._W = N - k$

$F = \dfrac{MS_{BET}}{MS_W}$ where $d.f.$ numerator $= d.f._{BET} = k - 1$;

$d.f.$ denominator $= d.f._W = N - k$

Two-Way ANOVA

r = number of rows; c = number of columns

Row factor F: $\dfrac{MS\ \text{row factor}}{MS\ \text{error}}$

Column factor F: $\dfrac{MS\ \text{column factor}}{MS\ \text{error}}$

Interaction F: $\dfrac{MS\ \text{interaction}}{MS\ \text{error}}$

with degrees of freedom for

row factor $= r - 1$ interaction $= (r - 1)(c - 1)$

olumn factor $= c - 1$ error $= rc(n - 1)$

Chapter 12

Sample test statistic for r = proportion of plus signs to all signs $(n \geq 12)$

$z = \dfrac{r - 0.5}{\sqrt{0.25/n}}$

Sample test statistic for R = sum of ranks

$z = \dfrac{R - \mu_R}{\sigma_R}$ where $\mu_R = \dfrac{n_1(n_1 + n_2 + 1)}{2}$ and

$\sigma_R = \sqrt{\dfrac{n_1 n_2 (n_1 + n_2 + 1)}{12}}$

Spearman rank correlation coefficient

$r_s = 1 - \dfrac{6\Sigma d^2}{n(n^2 - 1)}$ where $d = x - y$

1. Random Numbers
2. Binomial Coefficients $C_{n,r}$
3. Binomial Probability Distribution $C_{n,r}p^r q^{n-r}$
4. Poisson Probability Distribution
5. Areas of a Standard Normal Distribution

6. Student's t Distribution
7. Critical Values of Pearson Product-Moment Correlation, r
8. The χ^2 Distribution
9. The F Distribution
10. Critical Values for Spearman Rank Correlation, r_s

Table 1   Random Numbers

92630	78240	19267	95457	53497	23894	37708	79862	76471	66418
79445	78735	71549	44843	26104	67318	00701	34986	66751	99723
59654	71966	27386	50004	05358	94031	29281	18544	52429	06080
31524	49587	76612	39789	13537	48086	59483	60680	84675	53014
06348	76938	90379	51392	55887	71015	09209	79157	24440	30244
28703	51709	94456	48396	73780	06436	86641	69239	57662	80181
68108	89266	94730	95761	75023	48464	65544	96583	18911	16391
99938	90704	93621	66330	33393	95261	95349	51769	91616	33238
91543	73196	34449	63513	83834	99411	58826	40456	69268	48562
42103	02781	73920	56297	72678	12249	25270	36678	21313	75767
17138	27584	25296	28387	51350	61664	37893	05363	44143	42677
28297	14280	54524	21618	95320	38174	60579	08089	94999	78460
09331	56712	51333	06289	75345	08811	82711	57392	25252	30333
31295	04204	93712	51287	05754	79396	87399	51773	33075	97061
36146	15560	27592	42089	99281	59640	15221	96079	09961	05371
29553	18432	13630	05529	02791	81017	49027	79031	50912	09399
23501	22642	63081	08191	89420	67800	55137	54707	32945	64522
57888	85846	67967	07835	11314	01545	48535	17142	08552	67457
55336	71264	88472	04334	63919	36394	11196	92470	70543	29776
10087	10072	55980	64688	68239	20461	89381	93809	00796	95945
34101	81277	66090	88872	37818	72142	67140	50785	21380	16703
53362	44940	60430	22834	14130	96593	23298	56203	92671	15925
82975	66158	84731	19436	55790	69229	28661	13675	99318	76873
54827	84673	22898	08094	14326	87038	42892	21127	30712	48489
25464	59098	27436	89421	80754	89924	19097	67737	80368	08795
67609	60214	41475	84950	40133	02546	09570	45682	50165	15609
44921	70924	61295	51137	47596	86735	35561	76649	18217	63446
33170	30972	98130	95828	49786	13301	36081	80761	33985	68621
84687	85445	06208	17654	51333	02878	35010	67578	61574	20749
71886	56450	36567	09395	96951	35507	17555	35212	69106	01679

Table 1 continued

00475	02224	74722	14721	40215	21351	08596	45625	83981	63748
25993	38881	68361	59560	41274	69742	40703	37993	03435	18873
92882	53178	99195	93803	56985	53089	15305	50522	55900	43026
25138	26810	07093	15677	60688	04410	24505	37890	67186	62829
84631	71882	12991	83028	82484	90339	91950	74579	03539	90122
34003	92326	12793	61453	48121	74271	28363	66561	75220	35908
53775	45749	05734	86169	42762	70175	97310	73894	88606	19994
59316	97885	72807	54966	60859	11932	35265	71601	55577	67715
20479	66557	50705	26999	09854	52591	14063	30214	19890	19292
86180	84931	25455	26044	02227	52015	21820	50599	51671	65411
21451	68001	72710	40261	61281	13172	63819	48970	51732	54113
98062	68375	80089	24135	72355	95428	11808	29740	81644	86610
01788	64429	14430	94575	75153	94576	61393	96192	03227	32258
62465	04841	43272	68702	01274	05437	22953	18946	99053	41690
94324	31089	84159	92933	99989	89500	91586	02802	69471	68274
05797	43984	21575	09908	70221	19791	51578	36432	33494	79888
10395	14289	52185	09721	25789	38562	54794	04897	59012	89251
35177	56986	25549	59730	64718	52630	31100	62384	49483	11409
25633	89619	75882	98256	02126	72099	57183	55887	09320	73463
16464	48280	94254	45777	45150	68865	11382	11782	22695	41988

Source: Reprinted from *A Million Random Digits with 100,000 Normal Deviates* by the Rand Corporation (New York: The Free Press, 1955). Copyright 1955 and 1983 by the Rand Corporation. Used by permission.

Table 2   Binomial Coefficients C$_{n,r}$

n \ r	0	1	2	3	4	5	6	7	8	9	10
1	1	1									
2	1	2	1								
3	1	3	3	1							
4	1	4	6	4	1						
5	1	5	10	10	5	1					
6	1	6	15	20	15	6	1				
7	1	7	21	35	35	21	7	1			
8	1	8	28	56	70	56	28	8	1		
9	1	9	36	84	126	126	84	36	9	1	
10	1	10	45	120	210	252	210	120	45	10	1
11	1	11	55	165	330	462	462	330	165	55	11
12	1	12	66	220	495	792	924	792	495	220	66
13	1	13	78	286	715	1,287	1,716	1,716	1,287	715	286
14	1	14	91	364	1,001	2,002	3,003	3,432	3,003	2,002	1,001
15	1	15	105	455	1,365	3,003	5,005	6,435	6,435	5,005	3,003
16	1	16	120	560	1,820	4,368	8,008	11,440	12,870	11,440	8,008
17	1	17	136	680	2,380	6,188	12,376	19,448	24,310	24,310	19,448
18	1	18	153	816	3,060	8,568	18,564	31,824	43,758	48,620	43,758
19	1	19	171	969	3,876	11,628	27,132	50,388	75,582	92,378	92,378
20	1	20	190	1,140	4,845	15,504	38,760	77,520	125,970	167,960	184,756

Table 3 Binomial Probability Distribution $C_{n,r}\, p^r q^{n-r}$

This table shows the probability of r successes in n independent trials, each with probability of success p.

n	r	.01	.05	.10	.15	.20	.25	.30	.35	.40	.45	.50	.55	.60	.65	.70	.75	.80	.85	.90	.95
2	0	.980	.902	.810	.723	.640	.563	.490	.423	.360	.303	.250	.203	.160	.123	.090	.063	.040	.023	.010	.002
	1	.020	.095	.180	.255	.320	.375	.420	.455	.480	.495	.500	.495	.480	.455	.420	.375	.320	.255	.180	.095
	2	.000	.002	.010	.023	.040	.063	.090	.123	.160	.203	.250	.303	.360	.423	.490	.563	.640	.723	.810	.902
3	0	.970	.857	.729	.614	.512	.422	.343	.275	.216	.166	.125	.091	.064	.043	.027	.016	.008	.003	.001	.000
	1	.029	.135	.243	.325	.384	.422	.441	.444	.432	.408	.375	.334	.288	.239	.189	.141	.096	.057	.027	.007
	2	.000	.007	.027	.057	.096	.141	.189	.239	.288	.334	.375	.408	.432	.444	.441	.422	.384	.325	.243	.135
	3	.000	.000	.001	.003	.008	.016	.027	.043	.064	.091	.125	.166	.216	.275	.343	.422	.512	.614	.729	.857
4	0	.961	.815	.656	.522	.410	.316	.240	.179	.130	.092	.062	.041	.026	.015	.008	.004	.002	.001	.000	.000
	1	.039	.171	.292	.368	.410	.422	.412	.384	.346	.300	.250	.200	.154	.112	.076	.047	.026	.011	.004	.000
	2	.001	.014	.049	.098	.154	.211	.265	.311	.346	.368	.375	.368	.346	.311	.265	.211	.154	.098	.049	.014
	3	.000	.000	.004	.011	.026	.047	.076	.112	.154	.200	.250	.300	.346	.384	.412	.422	.410	.368	.292	.171
	4	.000	.000	.000	.001	.002	.004	.008	.015	.026	.041	.062	.092	.130	.179	.240	.316	.410	.522	.656	.815
5	0	.951	.774	.590	.444	.328	.237	.168	.116	.078	.050	.031	.019	.010	.005	.002	.001	.000	.000	.000	.000
	1	.048	.204	.328	.392	.410	.396	.360	.312	.259	.206	.156	.113	.077	.049	.028	.015	.006	.002	.000	.000
	2	.001	.021	.073	.138	.205	.264	.309	.336	.346	.337	.312	.276	.230	.181	.132	.088	.051	.024	.008	.001
	3	.000	.001	.008	.024	.051	.088	.132	.181	.230	.276	.312	.337	.346	.336	.309	.264	.205	.138	.073	.021
	4	.000	.000	.000	.002	.006	.015	.028	.049	.077	.113	.156	.206	.259	.312	.360	.396	.410	.392	.328	.204
	5	.000	.000	.000	.000	.000	.001	.002	.005	.010	.019	.031	.050	.078	.116	.168	.237	.328	.444	.590	.774
6	0	.941	.735	.531	.377	.262	.178	.118	.075	.047	.028	.016	.008	.004	.002	.001	.000	.000	.000	.000	.000
	1	.057	.232	.354	.399	.393	.356	.303	.244	.187	.136	.094	.061	.037	.020	.010	.004	.002	.000	.000	.000
	2	.001	.031	.098	.176	.246	.297	.324	.328	.311	.278	.234	.186	.138	.095	.060	.033	.015	.006	.001	.000
	3	.000	.002	.015	.042	.082	.132	.185	.236	.276	.303	.312	.303	.276	.236	.185	.132	.082	.042	.015	.002
	4	.000	.000	.001	.006	.015	.033	.060	.095	.138	.186	.234	.278	.311	.328	.324	.297	.246	.176	.098	.031
	5	.000	.000	.000	.000	.002	.004	.010	.020	.037	.061	.094	.136	.187	.244	.303	.356	.393	.399	.354	.232
	6	.000	.000	.000	.000	.000	.000	.001	.002	.004	.008	.016	.028	.047	.075	.118	.178	.262	.377	.531	.735
7	0	.932	.698	.478	.321	.210	.133	.082	.049	.028	.015	.008	.004	.002	.001	.000	.000	.000	.000	.000	.000
	1	.066	.257	.372	.396	.367	.311	.247	.185	.131	.087	.055	.032	.017	.008	.004	.001	.000	.000	.000	.000
	2	.002	.041	.124	.210	.275	.311	.318	.299	.261	.214	.164	.117	.077	.047	.025	.012	.004	.001	.000	.000
	3	.000	.004	.023	.062	.115	.173	.227	.268	.290	.292	.273	.239	.194	.144	.097	.058	.029	.011	.003	.000
	4	.000	.000	.003	.011	.029	.058	.097	.144	.194	.239	.273	.292	.290	.268	.227	.173	.115	.062	.023	.004
	5	.000	.000	.000	.001	.004	.012	.025	.047	.077	.117	.164	.214	.261	.299	.318	.311	.275	.210	.124	.041
	6	.000	.000	.000	.000	.000	.001	.004	.008	.017	.032	.055	.087	.131	.185	.247	.311	.367	.396	.372	.257
	7	.000	.000	.000	.000	.000	.000	.000	.001	.002	.004	.008	.015	.028	.049	.082	.133	.210	.321	.478	.698

Table 3 continued

n	r	.01	.05	.10	.15	.20	.25	.30	.35	.40	.45	.50	.55	.60	.65	.70	.75	.80	.85	.90	.95
8	0	.923	.663	.430	.272	.168	.100	.058	.032	.017	.008	.004	.002	.001	.000	.000	.000	.000	.000	.000	.000
	1	.075	.279	.383	.385	.336	.267	.198	.137	.090	.055	.031	.016	.008	.003	.001	.000	.000	.000	.000	.000
	2	.003	.051	.149	.238	.294	.311	.296	.259	.209	.157	.109	.070	.041	.022	.010	.004	.001	.000	.000	.000
	3	.000	.005	.033	.084	.147	.208	.254	.279	.279	.257	.219	.172	.124	.081	.047	.023	.009	.003	.000	.000
	4	.000	.000	.005	.018	.046	.087	.136	.188	.232	.263	.273	.263	.232	.188	.136	.087	.046	.018	.005	.000
	5	.000	.000	.000	.003	.009	.023	.047	.081	.124	.172	.219	.257	.279	.279	.254	.208	.147	.084	.033	.005
	6	.000	.000	.000	.000	.001	.004	.010	.022	.041	.070	.109	.157	.209	.259	.296	.311	.294	.238	.149	.051
	7	.000	.000	.000	.000	.000	.000	.001	.003	.008	.016	.031	.055	.090	.137	.198	.267	.336	.385	.383	.279
	8	.000	.000	.000	.000	.000	.000	.000	.000	.001	.002	.004	.008	.017	.032	.058	.100	.168	.272	.430	.663
9	0	.914	.630	.387	.232	.134	.075	.040	.021	.010	.005	.002	.001	.000	.000	.000	.000	.000	.000	.000	.000
	1	.083	.299	.387	.368	.302	.225	.156	.100	.060	.034	.018	.008	.004	.001	.000	.000	.000	.000	.000	.000
	2	.003	.063	.172	.260	.302	.300	.267	.216	.161	.111	.070	.041	.021	.010	.004	.001	.000	.000	.000	.000
	3	.000	.008	.045	.107	.176	.234	.267	.272	.251	.212	.164	.116	.074	.042	.021	.009	.003	.001	.000	.000
	4	.000	.001	.007	.028	.066	.117	.172	.219	.251	.260	.246	.213	.167	.118	.074	.039	.017	.005	.001	.000
	5	.000	.000	.001	.005	.017	.039	.074	.118	.167	.213	.246	.260	.251	.219	.172	.117	.066	.028	.007	.001
	6	.000	.000	.000	.001	.003	.009	.021	.042	.074	.116	.164	.212	.251	.272	.267	.234	.176	.107	.045	.008
	7	.000	.000	.000	.000	.000	.001	.004	.010	.021	.041	.070	.111	.161	.216	.267	.300	.302	.260	.172	.063
	8	.000	.000	.000	.000	.000	.000	.000	.001	.004	.008	.018	.034	.060	.100	.156	.225	.302	.368	.387	.299
	9	.000	.000	.000	.000	.000	.000	.000	.000	.000	.001	.002	.005	.010	.021	.040	.075	.134	.232	.387	.630
10	0	.904	.599	.349	.197	.107	.056	.028	.014	.006	.003	.001	.000	.000	.000	.000	.000	.000	.000	.000	.000
	1	.091	.315	.387	.347	.268	.188	.121	.072	.040	.021	.010	.004	.002	.000	.000	.000	.000	.000	.000	.000
	2	.004	.075	.194	.276	.302	.282	.233	.176	.121	.076	.044	.023	.011	.004	.001	.000	.000	.000	.000	.000
	3	.000	.010	.057	.130	.201	.250	.267	.252	.215	.166	.117	.075	.042	.021	.009	.003	.001	.000	.000	.000
	4	.000	.001	.011	.040	.088	.146	.200	.238	.251	.238	.205	.160	.111	.069	.037	.016	.006	.001	.000	.000
	5	.000	.000	.001	.008	.026	.058	.103	.154	.201	.234	.246	.234	.201	.154	.103	.058	.026	.008	.001	.000
	6	.000	.000	.000	.001	.006	.016	.037	.069	.111	.160	.205	.238	.251	.238	.200	.146	.088	.040	.011	.001
	7	.000	.000	.000	.000	.001	.003	.009	.021	.042	.075	.117	.166	.215	.252	.267	.250	.201	.130	.057	.010
	8	.000	.000	.000	.000	.000	.000	.001	.004	.011	.023	.044	.076	.121	.176	.233	.282	.302	.276	.194	.075
	9	.000	.000	.000	.000	.000	.000	.000	.000	.002	.004	.010	.021	.040	.072	.121	.188	.268	.347	.387	.315
	10	.000	.000	.000	.000	.000	.000	.000	.000	.000	.000	.001	.003	.006	.014	.028	.056	.107	.197	.349	.599
11	0	.895	.569	.314	.167	.086	.042	.020	.009	.004	.001	.000	.000	.000	.000	.000	.000	.000	.000	.000	.000
	1	.099	.329	.384	.325	.236	.155	.093	.052	.027	.013	.005	.002	.001	.000	.000	.000	.000	.000	.000	.000
	2	.005	.087	.213	.287	.295	.258	.200	.140	.089	.051	.027	.013	.005	.002	.001	.000	.000	.000	.000	.000
	3	.000	.014	.071	.152	.221	.258	.257	.225	.177	.126	.081	.046	.023	.010	.004	.001	.000	.000	.000	.000
	4	.000	.001	.016	.054	.111	.172	.220	.243	.236	.206	.161	.113	.070	.038	.017	.006	.002	.000	.000	.000
	5	.000	.000	.002	.013	.039	.080	.132	.183	.221	.236	.226	.193	.147	.099	.057	.027	.010	.002	.000	.000

Table 3 continued

n	r	.01	.05	.10	.15	.20	.25	.30	.35	.40	.45	.50	.55	.60	.65	.70	.75	.80	.85	.90	.95
11	6	.000	.000	.000	.002	.010	.027	.057	.099	.147	.193	.226	.236	.221	.183	.132	.080	.039	.013	.002	.000
	7	.000	.000	.000	.000	.002	.006	.017	.038	.070	.113	.161	.206	.236	.243	.220	.172	.111	.054	.016	.001
	8	.000	.000	.000	.000	.000	.001	.004	.010	.023	.046	.081	.126	.177	.225	.257	.258	.221	.152	.071	.014
	9	.000	.000	.000	.000	.000	.000	.001	.002	.005	.013	.027	.051	.089	.140	.200	.258	.295	.287	.213	.087
	10	.000	.000	.000	.000	.000	.000	.000	.000	.001	.002	.005	.013	.027	.052	.093	.155	.236	.325	.384	.329
	11	.000	.000	.000	.000	.000	.000	.000	.000	.000	.000	.000	.001	.004	.009	.020	.042	.086	.167	.314	.569
12	0	.886	.540	.282	.142	.069	.032	.014	.006	.002	.001	.000	.000	.000	.000	.000	.000	.000	.000	.000	.000
	1	.107	.341	.377	.301	.206	.127	.071	.037	.017	.008	.003	.001	.000	.000	.000	.000	.000	.000	.000	.000
	2	.006	.099	.230	.292	.283	.232	.168	.109	.064	.034	.016	.007	.002	.001	.000	.000	.000	.000	.000	.000
	3	.000	.017	.085	.172	.236	.258	.240	.195	.142	.092	.054	.028	.012	.005	.001	.000	.000	.000	.000	.000
	4	.000	.002	.021	.068	.133	.194	.231	.237	.213	.170	.121	.076	.042	.020	.008	.002	.001	.000	.000	.000
	5	.000	.000	.004	.019	.053	.103	.158	.204	.227	.223	.193	.149	.101	.059	.029	.011	.003	.001	.000	.000
	6	.000	.000	.000	.004	.016	.040	.079	.128	.177	.212	.226	.212	.177	.128	.079	.040	.016	.004	.000	.000
	7	.000	.000	.000	.001	.003	.011	.029	.059	.101	.149	.193	.223	.227	.204	.158	.103	.053	.019	.004	.000
	8	.000	.000	.000	.000	.001	.002	.008	.020	.042	.076	.121	.170	.213	.237	.231	.194	.133	.068	.021	.002
	9	.000	.000	.000	.000	.000	.000	.001	.005	.012	.028	.054	.092	.142	.195	.240	.258	.236	.172	.085	.017
	10	.000	.000	.000	.000	.000	.000	.000	.001	.002	.007	.016	.034	.064	.109	.168	.232	.283	.292	.230	.099
	11	.000	.000	.000	.000	.000	.000	.000	.000	.000	.001	.003	.008	.017	.037	.071	.127	.206	.301	.377	.341
	12	.000	.000	.000	.000	.000	.000	.000	.000	.000	.000	.000	.001	.002	.006	.014	.032	.069	.142	.282	.540
15	0	.860	.463	.206	.087	.035	.013	.005	.002	.000	.000	.000	.000	.000	.000	.000	.000	.000	.000	.000	.000
	1	.130	.366	.343	.231	.132	.067	.031	.013	.005	.002	.000	.000	.000	.000	.000	.000	.000	.000	.000	.000
	2	.009	.135	.267	.286	.231	.156	.092	.048	.022	.009	.003	.001	.000	.000	.000	.000	.000	.000	.000	.000
	3	.000	.031	.129	.218	.250	.225	.170	.111	.063	.032	.014	.005	.002	.000	.000	.000	.000	.000	.000	.000
	4	.000	.005	.043	.116	.188	.225	.219	.179	.127	.078	.042	.019	.007	.002	.001	.000	.000	.000	.000	.000
	5	.000	.001	.010	.045	.103	.165	.206	.212	.186	.140	.092	.051	.024	.010	.003	.001	.000	.000	.000	.000
	6	.000	.000	.002	.013	.043	.092	.147	.191	.207	.191	.153	.105	.061	.030	.012	.003	.001	.000	.000	.000
	7	.000	.000	.000	.003	.014	.039	.081	.132	.177	.201	.196	.165	.118	.071	.035	.013	.003	.001	.000	.000
	8	.000	.000	.000	.001	.003	.013	.035	.071	.118	.165	.196	.201	.177	.132	.081	.039	.014	.003	.000	.000
	9	.000	.000	.000	.000	.001	.003	.012	.030	.061	.105	.153	.191	.207	.191	.147	.092	.043	.013	.002	.000
	10	.000	.000	.000	.000	.000	.001	.003	.010	.024	.051	.092	.140	.186	.212	.206	.165	.103	.045	.010	.001
	11	.000	.000	.000	.000	.000	.000	.001	.002	.007	.019	.042	.078	.127	.179	.219	.225	.188	.116	.043	.005
	12	.000	.000	.000	.000	.000	.000	.000	.000	.002	.005	.014	.032	.063	.111	.170	.225	.250	.218	.129	.031
	13	.000	.000	.000	.000	.000	.000	.000	.000	.000	.001	.003	.009	.022	.048	.092	.156	.231	.286	.267	.135
	14	.000	.000	.000	.000	.000	.000	.000	.000	.000	.000	.000	.002	.005	.013	.031	.067	.132	.231	.343	.366
	15	.000	.000	.000	.000	.000	.000	.000	.000	.000	.000	.000	.000	.000	.002	.005	.013	.035	.087	.206	.463
16	0	.851	.440	.185	.074	.028	.010	.003	.001	.000	.000	.000	.000	.000	.000	.000	.000	.000	.000	.000	.000
	1	.138	.371	.329	.210	.113	.053	.023	.009	.003	.001	.000	.000	.000	.000	.000	.000	.000	.000	.000	.000

(Header over the probability columns: p)

Table 3 continued

n	r	.01	.05	.10	.15	.20	.25	.30	.35	.40	.45	.50	.55	.60	.65	.70	.75	.80	.85	.90	.95
16	2	.010	.146	.275	.277	.211	.134	.073	.035	.015	.006	.002	.001	.000	.000	.000	.000	.000	.000	.000	.000
	3	.000	.036	.142	.229	.246	.208	.146	.089	.047	.022	.009	.003	.001	.000	.000	.000	.000	.000	.000	.000
	4	.000	.006	.051	.131	.200	.225	.204	.155	.101	.057	.028	.011	.004	.001	.000	.000	.000	.000	.000	.000
	5	.000	.001	.014	.056	.120	.180	.210	.201	.162	.112	.067	.034	.014	.005	.001	.000	.000	.000	.000	.000
	6	.000	.000	.003	.018	.055	.110	.165	.198	.198	.168	.122	.075	.039	.017	.006	.001	.000	.000	.000	.000
	7	.000	.000	.000	.005	.020	.052	.101	.152	.189	.197	.175	.132	.084	.044	.019	.006	.001	.000	.000	.000
	8	.000	.000	.000	.001	.006	.020	.049	.092	.142	.181	.196	.181	.142	.092	.049	.020	.006	.001	.000	.000
	9	.000	.000	.000	.000	.001	.006	.019	.044	.084	.132	.175	.197	.189	.152	.101	.052	.020	.005	.000	.000
	10	.000	.000	.000	.000	.000	.001	.006	.017	.039	.075	.122	.168	.198	.198	.165	.110	.055	.018	.003	.000
	11	.000	.000	.000	.000	.000	.000	.001	.005	.014	.034	.067	.112	.162	.201	.210	.180	.120	.056	.014	.001
	12	.000	.000	.000	.000	.000	.000	.000	.001	.004	.011	.028	.057	.101	.155	.204	.225	.200	.131	.051	.006
	13	.000	.000	.000	.000	.000	.000	.000	.000	.001	.003	.009	.022	.047	.089	.146	.208	.246	.229	.142	.036
	14	.000	.000	.000	.000	.000	.000	.000	.000	.000	.001	.002	.006	.015	.035	.073	.134	.211	.277	.275	.146
	15	.000	.000	.000	.000	.000	.000	.000	.000	.000	.000	.000	.001	.003	.009	.023	.053	.113	.210	.329	.371
	16	.000	.000	.000	.000	.000	.000	.000	.000	.000	.000	.000	.000	.000	.001	.003	.010	.028	.074	.185	.440
20	0	.818	.358	.122	.039	.012	.003	.001	.000	.000	.000	.000	.000	.000	.000	.000	.000	.000	.000	.000	.000
	1	.165	.377	.270	.137	.058	.021	.007	.002	.000	.000	.000	.000	.000	.000	.000	.000	.000	.000	.000	.000
	2	.016	.189	.285	.229	.137	.067	.028	.010	.003	.001	.000	.000	.000	.000	.000	.000	.000	.000	.000	.000
	3	.001	.060	.190	.243	.205	.134	.072	.032	.012	.004	.001	.000	.000	.000	.000	.000	.000	.000	.000	.000
	4	.000	.013	.090	.182	.218	.190	.130	.074	.035	.014	.005	.001	.000	.000	.000	.000	.000	.000	.000	.000
	5	.000	.002	.032	.103	.175	.202	.179	.127	.075	.036	.015	.005	.001	.000	.000	.000	.000	.000	.000	.000
	6	.000	.000	.009	.045	.109	.169	.192	.171	.124	.075	.037	.015	.005	.001	.000	.000	.000	.000	.000	.000
	7	.000	.000	.002	.016	.055	.112	.164	.184	.166	.122	.074	.037	.015	.004	.001	.000	.000	.000	.000	.000
	8	.000	.000	.000	.005	.022	.061	.114	.161	.180	.162	.120	.073	.035	.014	.004	.001	.000	.000	.000	.000
	9	.000	.000	.000	.001	.007	.027	.065	.116	.160	.177	.160	.119	.071	.034	.012	.003	.000	.000	.000	.000
	10	.000	.000	.000	.000	.002	.010	.031	.069	.117	.159	.176	.159	.117	.069	.031	.010	.002	.000	.000	.000
	11	.000	.000	.000	.000	.000	.003	.012	.034	.071	.119	.160	.177	.160	.116	.065	.027	.007	.001	.000	.000
	12	.000	.000	.000	.000	.000	.001	.004	.014	.035	.073	.120	.162	.180	.161	.114	.061	.022	.005	.000	.000
	13	.000	.000	.000	.000	.000	.000	.001	.004	.015	.037	.074	.122	.166	.184	.164	.112	.055	.016	.002	.000
	14	.000	.000	.000	.000	.000	.000	.000	.001	.005	.015	.037	.075	.124	.171	.192	.169	.109	.045	.009	.000
	15	.000	.000	.000	.000	.000	.000	.000	.000	.001	.005	.015	.036	.075	.127	.179	.202	.175	.103	.032	.002
	16	.000	.000	.000	.000	.000	.000	.000	.000	.000	.001	.005	.014	.035	.074	.130	.190	.218	.182	.090	.013
	17	.000	.000	.000	.000	.000	.000	.000	.000	.000	.000	.001	.004	.012	.032	.072	.134	.205	.243	.190	.060
	18	.000	.000	.000	.000	.000	.000	.000	.000	.000	.000	.000	.001	.003	.010	.028	.067	.137	.229	.285	.189
	19	.000	.000	.000	.000	.000	.000	.000	.000	.000	.000	.000	.000	.000	.002	.007	.021	.058	.137	.270	.377
	20	.000	.000	.000	.000	.000	.000	.000	.000	.000	.000	.000	.000	.000	.000	.001	.003	.012	.039	.122	.358

Table 4 Poisson Probability Distribution

For a given value of λ, entry indicates the probability of obtaining a specified value of *r*.

					λ					
r	.1	.2	.3	.4	.5	.6	.7	.8	.9	1.0
0	.9048	.8187	.7408	.6703	.6065	.5488	.4966	.4493	.4066	.3679
1	.0905	.1637	.2222	.2681	.3033	.3293	.3476	.3595	.3659	.3679
2	.0045	.0164	.0333	.0536	.0758	.0988	.1217	.1438	.1647	.1839
3	.0002	.0011	.0033	.0072	.0126	.0198	.0284	.0383	.0494	.0613
4	.0000	.0001	.0003	.0007	.0016	.0030	.0050	.0077	.0111	.0153
5	.0000	.0000	.0000	.0001	.0002	.0004	.0007	.0012	.0020	.0031
6	.0000	.0000	.0000	.0000	.0000	.0000	.0001	.0002	.0003	.0005
7	.0000	.0000	.0000	.0000	.0000	.0000	.0000	.0000	.0000	.0001

					λ					
r	1.1	1.2	1.3	1.4	1.5	1.6	1.7	1.8	1.9	2.0
0	.3329	.3012	.2725	.2466	.2231	.2019	.1827	.1653	.1496	.1353
1	.3662	.3614	.3543	.3452	.3347	.3230	.3106	.2975	.2842	.2707
2	.2014	.2169	.2303	.2417	.2510	.2584	.2640	.2678	.2700	.2707
3	.0738	.0867	.0998	.1128	.1255	.1378	.1496	.1607	.1710	.1804
4	.0203	.0260	.0324	.0395	.0471	.0551	.0636	.0723	.0812	.0902
5	.0045	.0062	.0084	.0111	.0141	.0176	.0216	.0260	.0309	.0361
6	.0008	.0012	.0018	.0026	.0035	.0047	.0061	.0078	.0098	.0120
7	.0001	.0002	.0003	.0005	.0008	.0011	.0015	.0020	.0027	.0034
8	.0000	.0000	.0001	.0001	.0001	.0002	.0003	.0005	.0006	.0009
9	.0000	.0000	.0000	.0000	.0000	.0000	.0001	.0001	.0001	.0002

					λ					
r	2.1	2.2	2.3	2.4	2.5	2.6	2.7	2.8	2.9	3.0
0	.1225	.1108	.1003	.0907	.0821	.0743	.0672	.0608	.0550	.0498
1	.2572	.2438	.2306	.2177	.2052	.1931	.1815	.1703	.1596	.1494
2	.2700	.2681	.2652	.2613	.2565	.2510	.2450	.2384	.2314	.2240
3	.1890	.1966	.2033	.2090	.2138	.2176	.2205	.2225	.2237	.2240
4	.0992	.1082	.1169	.1254	.1336	.1414	.1488	.1557	.1622	.1680
5	.0417	.0476	.0538	.0602	.0668	.0735	.0804	.0872	.0940	.1008
6	.0146	.0174	.0206	.0241	.0278	.0319	.0362	.0407	.0455	.0504
7	.0044	.0055	.0068	.0083	.0099	.0118	.0139	.0163	.0188	.0216
8	.0011	.0015	.0019	.0025	.0031	.0038	.0047	.0057	.0068	.0081
9	.0003	.0004	.0005	.0007	.0009	.0011	.0014	.0018	.0022	.0027
10	.0001	.0001	.0001	.0002	.0002	.0003	.0004	.0005	.0006	.0008
11	.0000	.0000	.0000	.0000	.0000	.0001	.0001	.0001	.0002	.0002
12	.0000	.0000	.0000	.0000	.0000	.0000	.0000	.0000	.0000	.0001

Table 4 continued

					λ					
r	3.1	3.2	3.3	3.4	3.5	3.6	3.7	3.8	3.9	4.0
0	.0450	.0408	.0369	.0334	.0302	.0273	.0247	.0224	.0202	.0183
1	.1397	.1304	.1217	.1135	.1057	.0984	.0915	.0850	.0789	.0733
2	.2165	.2087	.2008	.1929	.1850	.1771	.1692	.1615	.1539	.1465
3	.2237	.2226	.2209	.2186	.2158	.2125	.2087	.2046	.2001	.1954
4	.1734	.1781	.1823	.1858	.1888	.1912	.1931	.1944	.1951	.1954
5	.1075	.1140	.1203	.1264	.1322	.1377	.1429	.1477	.1522	.1563
6	.0555	.0608	.0662	.0716	.0771	.0826	.0881	.0936	.0989	.1042
7	.0246	.2078	.0312	.0348	.0385	.0425	.0466	.0508	.0551	.0595
8	.0095	.0111	.0129	.0148	.0169	.0191	.0215	.0241	.0269	.0298
9	.0033	.0040	.0047	.0056	.0066	.0076	.0089	.0102	.0116	.0132
10	.0010	.0013	.0016	.0019	.0023	.0028	.0033	.0039	.0045	.0053
11	.0003	.0004	.0005	.0006	.0007	.0009	.0011	.0013	.0016	.0019
12	.0001	.0001	.0001	.0002	.0002	.0003	.0003	.0004	.0005	.0006
13	.0000	.0000	.0000	.0000	.0001	.0001	.0001	.0001	.0002	.0002
14	.0000	.0000	.0000	.0000	.0000	.0000	.0000	.0000	.0000	.0001

					λ					
r	4.1	4.2	4.3	4.4	4.5	4.6	4.7	4.8	4.9	5.0
0	.0166	.0150	.0136	.0123	.0111	.0101	.0091	.0082	.0074	.0067
1	.0679	.0630	.0583	.0540	.0500	.0462	.0427	.0395	.0365	.0337
2	.1393	.1323	.1254	.1188	.1125	.1063	.1005	.0948	.0894	.0842
3	.1904	.1852	.1798	.1743	.1687	.1631	.1574	.1517	.1460	.1404
4	.1951	.1944	.1933	.1917	.1898	.1875	.1849	.1820	.1789	.1755
5	.1600	.1633	.1662	.1687	.1708	.1725	.1738	.1747	.1753	.1755
6	.1093	.1143	.1191	.1237	.1281	.1323	.1362	.1398	.1432	.1462
7	.0640	.0686	.0732	.0778	.0824	.0869	.0914	.0959	.1002	.1044
8	.0328	.0360	.0393	.0428	.0463	.0500	.0537	.0575	.0614	.0653
9	.0150	.0168	.0188	.0209	.0232	.0255	.0280	.0307	.0334	.0363
10	.0061	.0071	.0081	.0092	.0104	.0118	.0132	.0147	.0164	.0181
11	.0023	.0027	.0032	.0037	.0043	.0049	.0056	.0064	.0073	.0082
12	.0008	.0009	.0011	.0014	.0016	.0019	.0022	.0026	.0030	.0034
13	.0002	.0003	.0004	.0005	.0006	.0007	.0008	.0009	.0011	.0013
14	.0001	.0001	.0001	.0001	.0002	.0002	.0003	.0003	.0004	.0005
15	.0000	.0000	.0000	.0000	.0001	.0001	.0001	.0001	.0001	.0002

Table 4 continued

					λ					
r	5.1	5.2	5.3	5.4	5.5	5.6	5.7	5.8	5.9	6.0
0	.0061	.0055	.0050	.0045	.0041	.0037	.0033	.0030	.0027	.0025
1	.0311	.0287	.0265	.0244	.0225	.0207	.0191	.0176	.0162	.0149
2	.0793	.0746	.0701	.0659	.0618	.0580	.0544	.0509	.0477	.0446
3	.1348	.1293	.1239	.1185	.1133	.1082	.1033	.0985	.0938	.0892
4	.1719	.1681	.1641	.1600	.1558	.1515	.1472	.1428	.1383	.1339
5	.1753	.1748	.1740	.1728	.1714	.1697	.1678	.1656	.1632	.1606
6	.1490	.1515	.1537	.1555	.1571	.1584	.1594	.1601	.1605	.1606
7	.1086	.1125	.1163	.1200	.1234	.1267	.1298	.1326	.1353	.1377
8	.0692	.0731	.0771	.0810	.0849	.0887	.0925	.0962	.0998	.1033
9	.0392	.0423	.0454	.0486	.0519	.0552	.0586	.0620	.0654	.0688
10	.0200	.0220	.0241	.0262	.0285	.0309	.0334	.0359	.0386	.0413
11	.0093	.0104	.0116	.0129	.0143	.0157	.0173	.0190	.0207	.0225
12	.0039	.0045	.0051	.0058	.0065	.0073	.0082	.0092	.0102	.0113
13	.0015	.0018	.0021	.0024	.0028	.0032	.0036	.0041	.0046	.0052
14	.0006	.0007	.0008	.0009	.0011	.0013	.0015	.0017	.0019	.0022
15	.0002	.0002	.0003	.0003	.0004	.0005	.0006	.0007	.0008	.0009
16	.0001	.0001	.0001	.0001	.0001	.0002	.0002	.0002	.0003	.0003
17	.0000	.0000	.0000	.0000	.0000	.0000	.0001	.0001	.0001	.0001

					λ					
r	6.1	6.2	6.3	6.4	6.5	6.6	6.7	6.8	6.9	7.0
0	.0022	.0020	.0018	.0017	.0015	.0014	.0012	.0011	.0010	.0009
1	.0137	.0126	.0116	.0106	.0098	.0090	.0082	.0076	.0070	.0064
2	.0417	.0390	.0364	.0340	.0318	.0296	.0276	.0258	.0240	.0223
3	.0848	.0806	.0765	.0726	.0688	.0652	.0617	.0584	.0552	.0521
4	.1294	.1249	.1205	.1162	.1118	.1076	.1034	.0992	.0952	.0912
5	.1579	.1549	.1519	.1487	.1454	.1420	.1385	.1349	.1314	.1277
6	.1605	.1601	.1595	.1586	.1575	.1562	.1546	.1529	.1511	.1490
7	.1399	.1418	.1435	.1450	.1462	.1472	.1480	.1486	.1489	.1490
8	.1066	.1099	.1130	.1160	.1188	.1215	.1240	.1263	.1284	.1304
9	.0723	.0757	.0791	.0825	.0858	.0891	.0923	.0954	.0985	.1014
10	.0441	.0469	.0498	.0528	.0558	.0588	.0618	.0649	.0679	.0710
11	.0245	.0265	.0285	.0307	.0330	.0353	.0377	.0401	.0426	.0452
12	.0124	.0137	.0150	.0164	.0179	.0194	.0210	.0227	.0245	.0264
13	.0058	.0065	.0073	.0081	.0089	.0098	.0108	.0119	.0130	.0142
14	.0025	.0029	.0033	.0037	.0041	.0046	.0052	.0058	.0064	.0071
15	.0010	.0012	.0014	.0016	.0018	.0020	.0023	.0026	.0029	.0033

Table 4 continued

					λ					
r	6.1	6.2	6.3	6.4	6.5	6.6	6.7	6.8	6.9	7.0
16	.0004	.0005	.0005	.0006	.0007	.0008	.0010	.0011	.0013	.0014
17	.0001	.0002	.0002	.0002	.0003	.0003	.0004	.0004	.0005	.0006
18	.0000	.0001	.0001	.0001	.0001	.0001	.0001	.0002	.0002	.0002
19	.0000	.0000	.0000	.0000	.0000	.0000	.0000	.0001	.0001	.0001

					λ					
r	7.1	7.2	7.3	7.4	7.5	7.6	7.7	7.8	7.9	8.0
0	.0008	.0007	.0007	.0006	.0006	.0005	.0005	.0004	.0004	.0003
1	.0059	.0054	.0049	.0045	.0041	.0038	.0035	.0032	.0029	.0027
2	.0208	.0194	.0180	.0167	.0156	.0145	.0134	.0125	.0116	.0107
3	.0492	.0464	.0438	.0413	.0389	.0366	.0345	.0324	.0305	.0286
4	.0874	.0836	.0799	.0764	.0729	.0696	.0663	.0632	.0602	.0573
5	.1241	.1204	.1167	.1130	.1094	.1057	.1021	.0986	.0951	.0916
6	.1468	.1445	.1420	.1394	.1367	.1339	.1311	.1282	.1252	.1221
7	.1489	.1486	.1481	.1474	.1465	.1454	.1442	.1428	.1413	.1396
8	.1321	.1337	.1351	.1363	.1373	.1382	.1388	.1392	.1395	.1396
9	.1042	.1070	.1096	.1121	.1144	.1167	.1187	.1207	.1224	.1241
10	.0740	.0770	.0800	.0829	.0858	.0887	.0914	.0941	.0967	.0993
11	.0478	.0504	.0531	.0558	.0585	.0613	.0640	.0667	.0695	.0722
12	.0283	.0303	.0323	.0344	.0366	.0388	.0411	.0434	.0457	.0481
13	.0154	.0168	.0181	.0196	.0211	.0227	.0243	.0260	.0278	.0296
14	.0078	.0086	.0095	.0104	.0113	.0123	.0134	.0145	.0157	.0169
15	.0037	.0041	.0046	.0051	.0057	.0062	.0069	.0075	.0083	.0090
16	.0016	.0019	.0021	.0024	.0026	.0030	.0033	.0037	.0041	.0045
17	.0007	.0008	.0009	.0010	.0012	.0013	.0015	.0017	.0019	.0021
18	.0003	.0003	.0004	.0004	.0005	.0006	.0006	.0007	.0008	.0009
19	.0001	.0001	.0001	.0002	.0002	.0002	.0003	.0003	.0003	.0004
20	.0000	.0000	.0001	.0001	.0001	.0001	.0001	.0001	.0001	.0002
21	.0000	.0000	.0000	.0000	.0000	.0000	.0000	.0000	.0001	.0001

					λ					
r	8.1	8.2	8.3	8.4	8.5	8.6	8.7	8.8	8.9	9.0
0	.0003	.0003	.0002	.0002	.0002	.0002	.0002	.0002	.0001	.0001
1	.0025	.0023	.0021	.0019	.0017	.0016	.0014	.0013	.0012	.0011
2	.0100	.0092	.0086	.0079	.0074	.0068	.0063	.0058	.0054	.0050
3	.0269	.0252	.0237	.0222	.0208	.0195	.0183	.0171	.0160	.0150

Table 4 continued

					λ					
r	8.1	8.2	8.3	8.4	8.5	8.6	8.7	8.8	8.9	9.0
4	.0544	.0517	.0491	.0466	.0443	.0420	.0398	.0377	.0357	.0337
5	.0882	.0849	.0816	.0784	.0752	.0722	.0692	.0663	.0635	.0607
6	.1191	.1160	.1128	.1097	.1066	.1034	.1003	.0972	.0941	.0911
7	.1378	.1358	.1338	.1317	.1294	.1271	.1247	.1222	.1197	.1171
8	.1395	.1392	.1388	.1382	.1375	.1366	.1356	.1344	.1332	.1318
9	.1256	.1269	.1280	.1290	.1299	.1306	.1311	.1315	.1317	.1318
10	.1017	.1040	.1063	.1084	.1104	.1123	.1140	.1157	.1172	.1186
11	.0749	.0776	.0802	.0828	.0853	.0878	.0902	.0925	.0948	.0970
12	.0505	.0530	.0555	.0579	.0604	.0629	.0654	.0679	.0703	.0728
13	.0315	.0334	.0354	.0374	.0395	.0416	.0438	.0459	.0481	.0504
14	.0182	.0196	.0210	.0225	.0240	.0256	.0272	.0289	.0306	.0324
15	.0098	.0107	.0116	.0126	.0136	.0147	.0158	.0169	.0182	.0194
16	.0050	.0055	.0060	.0066	.0072	.0079	.0086	.0093	.0101	.0109
17	.0024	.0026	.0029	.0033	.0036	.0040	.0044	.0048	.0053	.0058
18	.0011	.0012	.0014	.0015	.0017	.0019	.0021	.0024	.0026	.0029
19	.0005	.0005	.0006	.0007	.0008	.0009	.0010	.0011	.0012	.0014
20	.0002	.0002	.0002	.0003	.0003	.0004	.0004	.0005	.0005	.0006
21	.0001	.0001	.0001	.0001	.0001	.0002	.0002	.0002	.0002	.0003
22	.0000	.0000	.0000	.0000	.0001	.0001	.0001	.0001	.0001	.0001

					λ					
r	9.1	9.2	9.3	9.4	9.5	9.6	9.7	9.8	9.9	10
0	.0001	.0001	.0001	.0001	.0001	.0001	.0001	.0001	.0001	.0000
1	.0010	.0009	.0009	.0008	.0007	.0007	.0006	.0005	.0005	.0005
2	.0046	.0043	.0040	.0037	.0034	.0031	.0029	.0027	.0025	.0023
3	.0140	.0131	.0123	.0115	.0107	.0100	.0093	.0087	.0081	.0076
4	.0319	.0302	.0285	.0269	.0254	.0240	.0226	.0213	.0201	.0189
5	.0581	.0555	.0530	.0506	.0483	.0460	.0439	.0418	.0398	.0378
6	.0881	.0851	.0822	.0793	.0764	.0736	.0709	.0682	.0656	.0631
7	.1145	.1118	.1091	.1064	.1037	.1010	.0982	.0955	.0928	.0901
8	.1302	.1286	.1269	.1251	.1232	.1212	.1191	.1170	.1148	.1126
9	.1317	.1315	.1311	.1306	.1300	.1293	.1284	.1274	.1263	.1251
10	.1198	.1210	.1219	.1228	.1235	.1241	.1245	.1249	.1250	.1251
11	.0991	.1012	.1031	.1049	.1067	.1083	.1098	.1112	.1125	.1137
12	.0752	.0776	.0799	.0822	.0844	.0866	.0888	.0908	.0928	.0948
13	.0526	.0549	.0572	.0594	.0617	.0640	.0662	.0685	.0707	.0729
14	.0342	.0361	.0380	.0399	.0419	.0439	.0459	.0479	.0500	.0521

Table 4 continued

					λ					
r	9.1	9.2	9.3	9.4	9.5	9.6	9.7	9.8	9.9	10
15	.0208	.0221	.0235	.0250	.0265	.0281	.0297	.0313	.0330	.0347
16	.0118	.0127	.0137	.0147	.0157	.0168	.0180	.0192	.0204	.0217
17	.0063	.0069	.0075	.0081	.0088	.0095	.0103	.0111	.0119	.0128
18	.0032	.0035	.0039	.0042	.0046	.0051	.0055	.0060	.0065	.0071
19	.0015	.0017	.0019	.0021	.0023	.0026	.0028	.0031	.0034	.0037
20	.0007	.0008	.0009	.0010	.0011	.0012	.0014	.0015	.0017	.0019
21	.0003	.0003	.0004	.0004	.0005	.0006	.0006	.0007	.0008	.0009
22	.0001	.0001	.0002	.0002	.0002	.0002	.0003	.0003	.0004	.0004
23	.0000	.0001	.0001	.0001	.0001	.0001	.0001	.0001	.0002	.0002
24	.0000	.0000	.0000	.0000	.0000	.0000	.0000	.0001	.0001	.0001

					λ					
r	11	12	13	14	15	16	17	18	19	20
0	.0000	.0000	.0000	.0000	.0000	.0000	.0000	.0000	.0000	.0000
1	.0002	.0001	.0000	.0000	.0000	.0000	.0000	.0000	.0000	.0000
2	.0010	.0004	.0002	.0001	.0000	.0000	.0000	.0000	.0000	.0000
3	.0037	.0018	.0008	.0004	.0002	.0001	.0000	.0000	.0000	.0000
4	.0102	.0053	.0027	.0013	.0006	.0003	.0001	.0001	.0000	.0000
5	.0224	.0127	.0070	.0037	.0019	.0010	.0005	.0002	.0001	.0001
6	.0411	.0255	.0152	.0087	.0048	.0026	.0014	.0007	.0004	.0002
7	.0646	.0437	.0281	.0174	.0104	.0060	.0034	.0018	.0010	.0005
8	.0888	.0655	.0457	.0304	.0194	.0120	.0072	.0042	.0024	.0013
9	.1085	.0874	.0661	.0473	.0324	.0213	.0135	.0083	.0050	.0029
10	.1194	.1048	.0859	.0663	.0486	.0341	.0230	.0150	.0095	.0058
11	.1194	.1144	.1015	.0844	.0663	.0496	.0355	.0245	.0164	.0106
12	.1094	.1144	.1099	.0984	.0829	.0661	.0504	.0368	.0259	.0176
13	.0926	.1056	.1099	.1060	.0956	.0814	.0658	.0509	.0378	.0271
14	.0728	.0905	.1021	.1060	.1024	.0930	.0800	.0655	.0514	.0387
15	.0534	.0724	.0885	.0989	.1024	.0992	.0906	.0786	.0650	.0516
16	.0367	.0543	.0719	.0866	.0960	.0992	.0963	.0884	.0772	.0646
17	.0237	.0383	.0550	.0713	.0847	.0934	.0963	.0936	.0863	.0760
18	.0145	.0256	.0397	.0554	.0706	.0830	.0909	.0936	.0911	.0844
19	.0084	.0161	.0272	.0409	.0557	.0699	.0814	.0887	.0911	.0888
20	.0046	.0097	.0177	.0286	.0418	.0559	.0692	.0798	.0866	.0888
21	.0024	.0055	.0109	.0191	.0299	.0426	.0560	.0684	.0783	.0846

Table 4 continued

					λ					
r	11	12	13	14	15	16	17	18	19	20
22	.0012	.0030	.0065	.0121	.0204	.0310	.0433	.0560	.0676	.0769
23	.0006	.0016	.0037	.0074	.0133	.0216	.0320	.0438	.0559	.0669
24	.0003	.0008	.0020	.0043	.0083	.0144	.0226	.0328	.0442	.0557
25	.0001	.0004	.0010	.0024	.0050	.0092	.0154	.0237	.0336	.0446
26	.0000	.0002	.0005	.0013	.0029	.0057	.0101	.0164	.0246	.0343
27	.0000	.0001	.0002	.0007	.0016	.0034	.0063	.0109	.0173	.0254
28	.0000	.0000	.0001	.0003	.0009	.0019	.0038	.0070	.0117	.0181
29	.0000	.0000	.0001	.0002	.0004	.0011	.0023	.0044	.0077	.0125
30	.0000	.0000	.0000	.0001	.0002	.0006	.0013	.0026	.0049	.0083
31	.0000	.0000	.0000	.0000	.0001	.0003	.0007	.0015	.0030	.0054
32	.0000	.0000	.0000	.0000	.0001	.0001	.0004	.0009	.0018	.0034
33	.0000	.0000	.0000	.0000	.0000	.0001	.0002	.0005	.0010	.0020
34	.0000	.0000	.0000	.0000	.0000	.0000	.0001	.0002	.0006	.0012
35	.0000	.0000	.0000	.0000	.0000	.0000	.0000	.0001	.0003	.0007
36	.0000	.0000	.0000	.0000	.0000	.0000	.0000	.0001	.0002	.0004
37	.0000	.0000	.0000	.0000	.0000	.0000	.0000	.0000	.0001	.0002
38	.0000	.0000	.0000	.0000	.0000	.0000	.0000	.0000	.0000	.0001
39	.0000	.0000	.0000	.0000	.0000	.0000	.0000	.0000	.0000	.0001

Source: Extracted from William H. Beyer (ed.), *CRC Basic Statistical Tables* (Cleveland, Ohio: The Chemical Rubber Co., 1971).

Table 5 Areas of a Standard Normal Distribution

The table entries represent the area under the standard normal curve from 0 to the specified value of z.

z	.00	.01	.02	.03	.04	.05	.06	.07	.08	.09
0.0	.0000	.0040	.0080	.0120	.0160	.0199	.0239	.0279	.0319	.0359
0.1	.0398	.0438	.0478	.0517	.0557	.0596	.0636	.0675	.0714	.0753
0.2	.0793	.0832	.0871	.0910	.0948	.0987	.1026	.1064	.1103	.1141
0.3	.1179	.1217	.1255	.1293	.1331	.1368	.1406	.1443	.1480	.1517
0.4	.1554	.1591	.1628	.1664	.1700	.1736	.1772	.1808	.1844	.1879
0.5	.1915	.1950	.1985	.2019	.2054	.2088	.2123	.2157	.2190	.2224
0.6	.2257	.2291	.2324	.2357	.2389	.2422	.2454	.2486	.2517	.2549
0.7	.2580	.2611	.2642	.2673	.2704	.2734	.2764	.2794	.2823	.2852
0.8	.2881	.2910	.2939	.2967	.2995	.3023	.3051	.3078	.3106	.3133
0.9	.3159	.3186	.3212	.3238	.3264	.3289	.3315	.3340	.3365	.3389
1.0	.3413	.3438	.3461	.3485	.3508	.3531	.3554	.3577	.3599	.3621
1.1	.3643	.3665	.3686	.3708	.3729	.3749	.3770	.3790	.3810	.3830
1.2	.3849	.3869	.3888	.3907	.3925	.3944	.3962	.3980	.3997	.4015
1.3	.4032	.4049	.4066	.4082	.4099	.4115	.4131	.4147	.4162	.4177
1.4	.4192	.4207	.4222	.4236	.4251	.4265	.4279	.4292	.4306	.4319
1.5	.4332	.4345	.4357	.4370	.4382	.4394	.4406	.4418	.4429	.4441
1.6	.4452	.4463	.4474	.4484	.4495	.4505	.4515	.4525	.4535	.4545
1.7	.4554	.4564	.4573	.4582	.4591	.4599	.4608	.4616	.4625	.4633
1.8	.4641	.4649	.4656	.4664	.4671	.4678	.4686	.4693	.4699	.4706
1.9	.4713	.4719	.4726	.4732	.4738	.4744	.4750	.4756	.4761	.4767
2.0	.4772	.4778	.4783	.4788	.4793	.4798	.4803	.4808	.4812	.4817
2.1	.4821	.4826	.4830	.4834	.4838	.4842	.4846	.4850	.4854	.4857
2.2	.4861	.4864	.4868	.4871	.4875	.4878	.4881	.4884	.4887	.4890
2.3	.4893	.4896	.4898	.4901	.4904	.4906	.4909	.4911	.4913	.4916
2.4	.4918	.4920	.4922	.4925	.4927	.4929	.4931	.4932	.4934	.4936
2.5	.4938	.4940	.4941	.4943	.4945	.4946	.4948	.4949	.4951	.4952
2.6	.4953	.4955	.4956	.4957	.4959	.4960	.4961	.4962	.4963	.4964
2.7	.4965	.4966	.4967	.4968	.4969	.4970	.4971	.4972	.4973	.4974
2.8	.4974	.4975	.4976	.4977	.4977	.4978	.4979	.4979	.4980	.4981
2.9	.4981	.4982	.4982	.4983	.4984	.4984	.4985	.4985	.4986	.4986
3.0	.4987	.4987	.4987	.4988	.4988	.4989	.4989	.4989	.4990	.4990
3.1	.4990	.4991	.4991	.4991	.4992	.4992	.4992	.4992	.4993	.4993
3.2	.4993	.4993	.4994	.4994	.4994	.4994	.4994	.4995	.4995	.4995
3.3	.4995	.4995	.4995	.4996	.4996	.4996	.4996	.4996	.4996	.4997
3.4	.4997	.4997	.4997	.4997	.4997	.4997	.4997	.4997	.4997	.4998
3.5	.4998	.4998	.4998	.4998	.4998	.4998	.4998	.4998	.4998	.4998
3.6	.4998	.4998	.4998	.4999	.4999	.4999	.4999	.4999	.4999	.4999

For values of z greater than or equal 3.70, use 0.4999 to approximate the shaded area under the standard normal curve.

Table 6 Student's t Distribution

	c	0.750	0.800	0.850	0.900	0.950	0.980	0.990
	α'	0.125	0.100	0.075	0.050	0.025	0.010	0.005
d.f.	α''	0.250	0.200	0.150	0.100	0.050	0.020	0.010
1		2.414	3.078	4.165	6.314	12.706	31.821	63.657
2		1.604	1.886	2.282	2.920	4.303	6.965	9.925
3		1.423	1.638	1.924	2.353	3.182	4.541	5.841
4		1.344	1.533	1.778	2.132	2.776	3.747	4.604
5		1.301	1.476	1.699	2.015	2.571	3.365	4.032
6		1.273	1.440	1.650	1.943	2.447	3.143	3.707
7		1.254	1.415	1.617	1.895	2.365	2.998	3.499
8		1.240	1.397	1.592	1.860	2.306	2.896	3.355
9		1.230	1.383	1.574	1.833	2.262	2.821	3.250
10		1.221	1.372	1.559	1.812	2.228	2.764	3.169
11		1.214	1.363	1.548	1.796	2.201	2.718	3.106
12		1.209	1.356	1.538	1.782	2.179	2.681	3.055
13		1.204	1.350	1.530	1.771	2.160	2.650	3.012
14		1.200	1.345	1.523	1.761	2.145	2.624	2.977
15		1.197	1.341	1.517	1.753	2.131	2.602	2.947
16		1.194	1.337	1.512	1.746	2.120	2.583	2.921
17		1.191	1.333	1.508	1.740	2.110	2.567	2.898
18		1.189	1.330	1.504	1.734	2.101	2.552	2.878
19		1.187	1.328	1.500	1.729	2.093	2.539	2.861
20		1.185	1.325	1.497	1.725	2.086	2.528	2.845
21		1.183	1.323	1.494	1.721	2.080	2.518	2.831
22		1.182	1.321	1.492	1.717	2.074	2.508	2.819
23		1.180	1.319	1.489	1.714	2.069	2.500	2.807
24		1.179	1.318	1.487	1.711	2.064	2.492	2.797
25		1.178	1.316	1.485	1.708	2.060	2.485	2.787
26		1.177	1.315	1.483	1.706	2.056	2.479	2.779
27		1.176	1.314	1.482	1.703	2.052	2.473	2.771
28		1.175	1.313	1.480	1.701	2.048	2.467	2.763
29		1.174	1.311	1.479	1.699	2.045	2.462	2.756
30		1.173	1.310	1.477	1.697	2.042	2.457	2.750
35		1.170	1.306	1.472	1.690	2.030	2.438	2.724
40		1.167	1.303	1.468	1.684	2.021	2.423	2.704
45		1.165	1.301	1.465	1.679	2.014	2.412	2.690
50		1.164	1.299	1.462	1.676	2.009	2.403	2.678
55		1.163	1.297	1.460	1.673	2.004	2.396	2.668
60		1.162	1.296	1.458	1.671	2.000	2.390	2.660
90		1.158	1.291	1.452	1.662	1.987	2.369	2.632
120		1.156	1.289	1.449	1.658	1.980	2.358	2.617
∞		1.150	1.282	1.440	1.645	1.960	2.326	2.58

c is a confidence level:

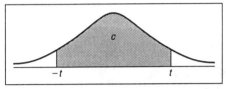

α' is the level of significance for a one-tailed test:

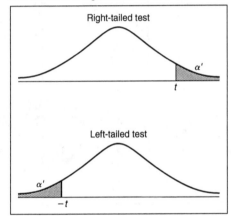

α'' is the level of significance for a two-tailed test:

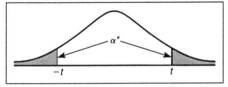

Table 7 Critical Values of Pearson Product Moment Correlation Coefficient, *r*

n	α = 0.01 one tail	α = 0.01 two tails	α = 0.05 one tail	α = 0.05 two tails
3	1.00	1.00	.99	1.00
4	.98	.99	.90	.95
5	.93	.96	.81	.88
6	.88	.92	.73	.81
7	.83	.87	.67	.75
8	.79	.83	.62	.71
9	.75	.80	.58	.67
10	.72	.76	.54	.63
11	.69	.73	.52	.60
12	.66	.71	.50	.58
13	.63	.68	.48	.53
14	.61	.66	.46	.53
15	.59	.64	.44	.51
16	.57	.61	.42	.50
17	.56	.61	.41	.48
18	.54	.59	.40	.47
19	.53	.58	.39	.46
20	.52	.56	.38	.44
21	.50	.55	.37	.43
22	.49	.54	.36	.42
23	.48	.53	.35	.41
24	.47	.52	.34	.40
25	.46	.51	.34	.40
26	.45	.50	.33	.39
27	.45	.49	.32	.38
28	.44	.48	.32	.37
29	.43	.47	.31	.37
30	.42	.46	.31	.36

For a right-tailed test, use a positive *r* value:

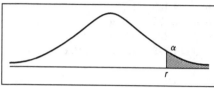

For a left-tailed test, use a negative *r* value:

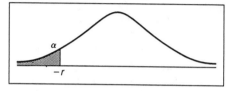

For a two-tailed test, use a positive *r* value and negative *r* value:

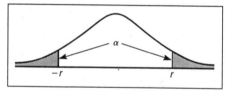

Table 8 The χ^2 Distribution

For d.f. ≥ 3

For d.f. = 1 or 2

d.f.\α	.995	.990	.975	.950	.900	.100	.050	.025	.010	.005
1	0.0^4393	0.0^3157	0.0^3982	0.0^2393	0.0158	2.71	3.84	5.02	6.63	7.88
2	0.0100	0.0201	0.0506	0.103	0.211	4.61	5.99	7.38	9.21	10.60
3	0.072	0.115	0.216	0.352	0.584	6.25	7.81	9.35	11.34	12.84
4	0.207	0.297	0.484	0.711	1.064	7.78	9.49	11.14	13.28	14.86
5	0.412	0.554	0.831	1.145	1.61	9.24	11.07	12.83	15.09	16.75
6	0.676	0.872	1.24	1.64	2.20	10.64	12.59	14.45	16.81	18.55
7	0.989	1.24	1.69	2.17	2.83	12.02	14.07	16.01	18.48	20.28
8	1.34	1.65	2.18	2.73	3.49	13.36	15.51	17.53	20.09	21.96
9	1.73	2.09	2.70	3.33	4.17	14.68	16.92	19.02	21.67	23.59
10	2.16	2.56	3.25	3.94	4.87	15.99	18.31	20.48	23.21	25.19
11	2.60	3.05	3.82	4.57	5.58	17.28	19.68	21.92	24.72	26.76
12	3.07	3.57	4.40	5.23	6.30	18.55	21.03	23.34	26.22	28.30
13	3.57	4.11	5.01	5.89	7.04	19.81	22.36	24.74	27.69	29.82
14	4.07	4.66	5.63	6.57	7.79	21.06	23.68	26.12	29.14	31.32
15	4.60	5.23	6.26	7.26	8.55	22.31	25.00	27.49	30.58	32.80
16	5.14	5.81	6.91	7.96	9.31	23.54	26.30	28.85	32.00	34.27
17	5.70	6.41	7.56	8.67	10.09	24.77	27.59	30.19	33.41	35.72
18	6.26	7.01	8.23	9.39	10.86	25.99	28.87	31.53	34.81	37.16
19	6.84	7.63	8.91	10.12	11.65	27.20	30.14	32.85	36.19	38.58
20	7.43	8.26	8.59	10.85	12.44	28.41	31.41	34.17	37.57	40.00
21	8.03	8.90	10.28	11.59	13.24	29.62	32.67	35.48	38.93	41.40
22	8.64	9.54	10.98	12.34	14.04	30.81	33.92	36.78	40.29	42.80
23	9.26	10.20	11.69	13.09	14.85	32.01	35.17	38.08	41.64	44.18
24	9.89	10.86	12.40	13.85	15.66	33.20	36.42	39.36	42.98	45.56
25	10.52	11.52	13.12	14.61	16.47	34.38	37.65	40.65	44.31	46.93
26	11.16	12.20	13.84	15.38	17.29	35.56	38.89	41.92	45.64	48.29
27	11.81	12.88	14.57	16.15	18.11	36.74	40.11	43.19	46.96	49.64
28	12.46	13.56	15.31	16.93	18.94	37.92	41.34	44.46	48.28	50.99
29	13.21	14.26	16.05	17.71	19.77	39.09	42.56	45.72	49.59	52.34
30	13.79	14.95	16.79	18.49	20.60	40.26	43.77	46.98	50.89	53.67
40	20.71	22.16	24.43	26.51	29.05	51.80	55.76	59.34	63.69	66.77
50	27.99	29.71	32.36	34.76	37.69	63.17	67.50	71.42	76.15	79.49
60	35.53	37.48	40.48	43.19	46.46	74.40	79.08	83.30	88.38	91.95
70	43.28	45.44	48.76	51.74	55.33	85.53	90.53	95.02	100.4	104.2
80	51.17	53.54	57.15	60.39	64.28	96.58	101.9	106.6	112.3	116.3
90	59.20	61.75	65.65	69.13	73.29	107.6	113.1	118.1	124.1	128.3
100	67.33	70.06	74.22	77.93	82.36	118.5	124.3	129.6	135.8	140.2

Source: From H. L. Herter, *Biometrika,* June 1964. Printed by permission of Biometrika Trustees.

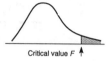

Critical value F ↑

Table 9 The *F* Distribution

Degrees of Freedom for Denominator	α	Degrees of Freedom for Numerator								
		1	2	3	4	5	6	7	8	9
1	.050	161.45	199.50	215.71	224.58	230.16	233.99	236.77	238.88	240.54
1	.025	647.79	799.50	864.16	899.58	921.85	937.11	948.22	956.66	963.28
1	.010	4052.2	4999.5	5403.4	5624.6	5763.6	5859.0	5928.4	5981.1	6022.5
2	.050	18.51	19.00	19.16	19.25	19.30	19.33	19.35	19.37	19.38
2	.025	38.51	39.00	39.17	39.25	39.30	39.33	39.36	39.37	39.39
2	.010	98.50	99.00	99.17	99.25	99.30	99.33	99.36	99.37	99.39
3	.050	10.13	9.55	9.28	9.12	9.01	8.94	8.89	8.85	8.81
3	.025	17.44	16.04	15.44	15.10	14.88	14.73	14.62	14.54	14.47
3	.010	34.12	30.82	29.46	28.71	28.24	27.91	27.67	27.49	27.35
4	.050	7.71	6.94	6.59	6.39	6.26	6.16	6.09	6.04	6.00
4	.025	12.22	10.65	9.98	9.60	9.36	9.20	9.07	8.98	8.90
4	.010	21.20	18.00	16.69	15.98	15.52	15.21	14.98	14.80	14.66
5	.050	6.61	5.79	5.41	5.19	5.05	4.95	4.88	4.82	4.77
5	.025	10.01	8.43	7.76	7.39	7.15	6.98	6.85	6.76	6.68
5	.010	16.26	13.27	12.06	11.39	10.97	10.67	10.46	10.29	10.16
6	.050	5.99	5.14	4.76	4.53	4.39	4.28	4.21	4.15	4.10
6	.025	8.81	7.26	6.60	6.23	5.99	5.82	5.70	5.60	5.52
6	.010	13.75	10.92	9.78	9.15	8.75	8.47	8.26	8.10	7.98
7	.050	5.59	4.74	4.35	4.12	3.97	3.87	3.79	3.73	3.68
7	.025	8.07	6.54	5.89	5.52	5.29	5.12	4.99	4.90	4.82
7	.010	12.25	9.55	8.45	7.85	7.46	7.19	6.99	6.84	6.72
8	.050	5.32	4.46	4.07	3.84	3.69	3.58	3.50	3.44	3.39
8	.025	7.57	6.06	5.42	5.05	4.82	4.65	4.53	4.43	4.36
8	.010	11.26	8.65	7.59	7.01	6.63	6.37	6.18	6.03	5.91
9	.050	5.12	4.26	3.86	3.63	3.48	3.37	3.29	3.23	3.18
9	.025	7.21	5.71	5.08	4.72	4.48	4.32	4.20	4.10	4.03
9	.010	10.56	8.02	6.99	6.42	6.06	5.80	5.61	5.47	5.35
10	.050	4.96	4.10	3.71	3.48	3.33	3.22	3.14	3.07	3.02
10	.025	6.94	5.46	4.83	4.47	4.24	4.07	3.95	3.85	3.78
10	.010	10.04	7.56	6.55	5.99	5.64	5.39	5.20	5.06	4.94
11	.050	4.84	3.98	3.59	3.36	3.20	3.09	3.01	2.95	2.90
11	.025	6.72	5.26	4.63	4.28	4.04	3.88	3.76	3.66	3.59
11	.010	9.65	7.21	6.22	5.67	5.32	5.07	4.89	4.74	4.63
12	.050	4.75	3.89	3.49	3.26	3.11	3.00	2.91	2.85	2.80
12	.025	6.55	5.10	4.47	4.12	3.89	3.73	3.61	3.51	3.44
12	.010	9.33	6.93	5.95	5.41	5.06	4.82	4.64	4.50	4.39

Table 9 continued

			Degrees of Freedom for Numerator							
10	12	15	20	25	30	40	50	60	120	1000
241.88	243.91	245.95	248.01	249.26	250.10	251.14	251.77	252.20	253.25	254.19
968.63	976.71	984.87	993.10	998.08	1001.4	1005.6	1008.1	1009.8	1014.0	1017.7
6055.8	6106.3	6157.3	6208.7	6239.8	6260.6	6286.8	6302.5	6313.0	6339.4	6362.7
19.40	19.41	19.43	19.45	19.46	19.46	19.47	19.48	19.48	19.49	19.49
39.40	39.41	39.43	39.45	39.46	39.46	39.47	39.48	39.48	39.49	39.50
99.40	99.42	99.43	99.45	99.46	99.47	99.47	99.48	99.48	99.49	99.50
8.79	8.74	8.70	8.66	8.63	8.62	8.59	8.58	8.57	8.55	8.53
14.42	14.34	14.25	14.17	14.12	14.08	14.04	14.01	13.99	13.95	13.91
27.23	27.05	26.87	26.69	26.58	26.50	26.41	26.35	26.32	26.22	26.14
5.96	5.91	5.86	5.80	5.77	5.75	5.72	5.70	5.69	5.66	5.63
8.84	8.75	8.66	8.56	8.50	8.46	8.41	8.38	8.36	8.31	8.26
14.55	14.37	14.20	14.02	13.91	13.84	13.75	13.69	13.65	13.56	13.47
4.74	4.68	4.62	4.56	4.52	4.50	4.46	4.44	4.43	4.40	4.37
6.62	6.52	6.43	6.33	6.27	6.23	6.18	6.14	6.12	6.07	6.02
10.05	9.89	9.72	9.55	9.45	9.38	9.29	9.24	9.20	9.11	9.03
4.06	4.00	3.94	3.87	3.83	3.81	3.77	3.75	3.74	3.70	3.67
5.46	5.37	5.27	5.17	5.11	5.07	5.01	4.98	4.96	4.90	4.86
7.87	7.72	7.56	7.40	7.30	7.23	7.14	7.09	7.06	6.97	6.89
3.64	3.57	3.51	3.44	3.40	3.38	3.34	3.32	3.30	3.27	3.23
4.76	4.67	4.57	4.47	4.40	4.36	4.31	4.28	4.25	4.20	4.15
6.62	6.47	6.31	6.16	6.06	5.99	5.91	5.86	5.82	5.74	5.66
3.35	3.28	3.22	3.15	3.11	3.08	3.04	3.02	3.01	2.97	2.93
4.30	4.20	4.10	4.00	3.94	3.89	3.84	3.81	3.78	3.73	3.68
5.81	5.67	5.52	5.36	5.26	5.20	5.12	5.07	5.03	4.95	4.87
3.14	3.07	3.01	2.94	2.89	2.86	2.83	2.80	2.79	2.75	2.71
3.96	3.87	3.77	3.67	3.60	3.56	3.51	3.47	3.45	3.39	3.34
5.26	5.11	4.96	4.81	4.71	4.65	4.57	4.52	4.48	4.40	4.32
2.98	2.91	2.85	2.77	2.73	2.70	2.66	2.64	2.62	2.58	2.54
3.72	3.62	3.52	3.42	3.35	3.31	3.26	3.22	3.20	3.14	3.09
4.85	4.71	4.56	4.41	4.31	4.25	4.17	4.12	4.08	4.00	3.92
2.85	2.79	2.72	2.65	2.60	2.57	2.53	2.51	2.49	2.45	2.41
3.53	3.43	3.33	3.23	3.16	3.12	3.06	3.03	3.00	2.94	2.89
4.54	4.40	4.25	4.10	4.01	3.94	3.86	3.81	3.78	3.69	3.61
2.75	2.69	2.62	2.54	2.50	2.47	2.43	2.40	2.38	2.34	2.30
3.37	3.28	3.18	3.07	3.01	2.96	2.91	2.87	2.85	2.79	2.73
4.30	4.16	4.01	3.86	3.76	3.70	3.62	3.57	3.54	3.45	3.37

Table 9 continued

Degrees of Freedom for Denominator	α	Degrees of Freedom for Numerator								
		1	2	3	4	5	6	7	8	9
13	0.050	4.67	3.81	3.41	3.18	3.03	2.92	2.83	2.77	2.71
13	0.025	6.41	4.97	4.35	4.00	3.77	3.60	3.48	3.39	3.31
13	0.010	9.07	6.70	5.74	5.21	4.86	4.62	4.44	4.30	4.19
14	0.050	4.60	3.74	3.34	3.11	2.96	2.85	2.76	2.70	2.65
14	0.025	6.30	4.86	4.24	3.89	3.66	3.50	3.38	3.29	3.21
14	0.010	8.86	6.51	5.56	5.04	4.69	4.46	4.28	4.14	4.03
15	0.050	4.54	3.68	3.29	3.06	2.90	2.79	2.71	2.64	2.59
15	0.025	6.20	4.77	4.15	3.80	3.58	3.41	3.29	3.20	3.12
15	0.010	8.68	6.36	5.42	4.89	4.56	4.32	4.14	4.00	3.89
16	0.050	4.49	3.63	3.24	3.01	2.85	2.74	2.66	2.59	2.54
16	0.025	6.12	4.69	4.08	3.73	3.50	3.34	3.22	3.12	3.05
16	0.010	8.53	6.23	5.29	4.77	4.44	4.20	4.03	3.89	3.78
17	0.050	4.45	3.59	3.20	2.96	2.81	2.70	2.61	2.55	2.49
17	0.025	6.04	4.62	4.01	3.66	3.44	3.28	3.16	3.06	2.98
17	0.010	8.40	6.11	5.19	4.67	4.34	4.10	3.93	3.79	3.68
18	0.050	4.41	3.55	3.16	2.93	2.77	2.66	2.58	2.51	2.46
18	0.025	5.98	4.56	3.95	3.61	3.38	3.22	3.10	3.01	2.93
18	0.010	8.29	6.01	5.09	4.58	4.25	4.01	3.84	3.71	3.60
19	0.050	4.38	3.52	3.13	2.90	2.74	2.63	2.54	2.48	2.42
19	0.025	5.92	4.51	3.90	3.56	3.33	3.17	3.05	2.96	2.88
19	0.010	8.18	5.93	5.01	4.50	4.17	3.94	3.77	3.63	3.52
20	0.050	4.35	3.49	3.10	2.87	2.71	2.60	2.51	2.45	2.39
20	0.025	5.87	4.46	3.86	3.51	3.29	3.13	3.01	2.91	2.84
20	0.010	8.10	5.85	4.94	4.43	4.10	3.87	3.70	3.56	3.46
21	0.050	4.32	3.47	3.07	2.84	2.68	2.57	2.49	2.42	2.37
21	0.025	5.83	4.42	3.82	3.48	3.25	3.09	2.97	2.87	2.80
21	0.010	8.02	5.78	4.87	4.37	4.04	3.81	3.64	3.51	3.40
22	0.050	4.30	3.44	3.05	2.82	2.66	2.55	2.46	2.40	2.34
22	0.025	5.79	4.38	3.78	3.44	3.22	3.05	2.93	2.84	2.76
22	0.010	7.95	5.72	4.82	4.31	3.99	3.76	3.59	3.45	3.35
23	0.050	4.28	3.42	3.03	2.80	2.64	2.53	2.44	2.37	2.32
23	0.025	5.75	4.35	3.75	3.41	3.18	3.02	2.90	2.81	2.73
23	0.010	7.88	5.66	4.76	4.26	3.94	3.71	3.54	3.41	3.30
24	0.050	4.26	3.40	3.01	2.78	2.62	2.51	2.42	2.36	2.30
24	0.025	5.72	4.32	3.72	3.38	3.15	2.99	2.87	2.78	2.70
24	0.010	7.82	5.61	4.72	4.22	3.90	3.67	3.50	3.36	3.26

Table 9 continued

			Degrees of Freedom for Numerator							
10	12	15	20	25	30	40	50	60	120	1000
2.67	2.60	2.53	2.46	2.41	2.38	2.34	2.31	2.30	2.25	2.21
3.25	3.15	3.05	2.95	2.88	2.84	2.78	2.74	2.72	2.66	2.60
4.10	3.96	3.82	3.66	3.57	3.51	3.43	3.38	3.34	3.25	3.18
2.60	2.53	2.46	2.39	2.34	2.31	2.27	2.24	2.22	2.18	2.14
3.15	3.05	2.95	2.84	2.78	2.73	2.67	2.64	2.61	2.55	2.50
3.94	3.80	3.66	3.51	3.41	3.35	3.27	3.22	3.18	3.09	3.02
2.54	2.48	2.40	2.33	2.28	2.25	2.20	2.18	2.16	2.11	2.07
3.06	2.96	2.86	2.76	2.69	2.64	2.59	2.55	2.52	2.46	2.40
3.80	3.67	3.52	3.37	3.28	3.21	3.13	3.08	3.05	2.96	2.88
2.49	2.42	2.35	2.28	2.23	2.19	2.15	2.12	2.11	2.06	2.02
2.99	2.89	2.79	2.68	2.61	2.57	2.51	2.47	2.45	2.38	2.32
3.69	3.55	3.41	3.26	3.16	3.10	3.02	2.97	2.93	2.84	2.76
2.45	2.38	2.31	2.23	2.18	2.15	2.10	2.08	2.06	2.01	1.97
2.92	2.82	2.72	2.62	2.55	2.50	2.44	2.41	2.38	2.32	2.26
3.59	3.46	3.31	3.16	3.07	3.00	2.92	2.87	2.83	2.75	2.66
2.41	2.34	2.27	2.19	2.14	2.11	2.06	2.04	2.02	1.97	1.92
2.87	2.77	2.67	2.56	2.49	2.44	2.38	2.35	2.32	2.26	2.20
3.51	3.37	3.23	3.08	2.98	2.92	2.84	2.78	2.75	2.66	2.58
2.38	2.31	2.23	2.16	2.11	2.07	2.03	2.00	1.98	1.93	1.88
2.82	2.72	2.62	2.51	2.44	2.39	2.33	2.30	2.27	2.20	2.14
3.43	3.30	3.15	3.00	2.91	2.84	2.76	2.71	2.67	2.58	2.50
2.35	2.28	2.20	2.12	2.07	2.04	1.99	1.97	1.95	1.90	1.85
2.77	2.68	2.57	2.46	2.40	2.35	2.29	2.25	2.22	2.16	2.09
3.37	3.23	3.09	2.94	2.84	2.78	2.69	2.64	2.61	2.52	2.43
2.32	2.25	2.18	2.10	2.05	2.01	1.96	1.94	1.92	1.87	1.82
2.73	2.64	2.53	2.42	2.36	2.31	2.25	2.21	2.18	2.11	2.05
3.31	3.17	3.03	2.88	2.79	2.72	2.64	2.58	2.55	2.46	2.37
2.30	2.23	2.15	2.07	2.02	1.98	1.94	1.91	1.89	1.84	1.79
2.70	2.60	2.50	2.39	2.32	2.27	2.21	2.17	2.14	2.08	2.01
3.26	3.12	2.98	2.83	2.73	2.67	2.58	2.53	2.50	2.40	2.32
2.27	2.20	2.13	2.05	2.00	1.96	1.91	1.88	1.86	1.81	1.76
2.67	2.57	2.47	2.36	2.29	2.24	2.18	2.14	2.11	2.04	1.98
3.21	3.07	2.93	2.78	2.69	2.62	2.54	2.48	2.45	2.35	2.27
2.25	2.18	2.11	2.03	1.97	1.94	1.89	1.86	1.84	1.79	1.74
2.64	2.54	2.44	2.33	2.26	2.21	2.15	2.11	2.08	2.01	1.94
3.17	3.03	2.89	2.74	2.64	2.58	2.49	2.44	2.40	2.31	2.22

Table 9 continued

Degrees of Freedom for Denominator	α	Degrees of Freedom for Numerator								
		1	2	3	4	5	6	7	8	9
25	0.05	4.24	3.39	2.99	2.76	2.60	2.49	2.40	2.34	2.28
25	0.025	5.69	4.29	3.69	3.35	3.13	2.97	2.85	2.75	2.68
25	0.01	7.77	5.57	4.68	4.18	3.85	3.63	3.46	3.32	3.22
26	0.05	4.23	3.37	2.98	2.74	2.59	2.47	2.39	2.32	2.27
26	0.025	5.66	4.27	3.67	3.33	3.10	2.94	2.82	2.73	2.65
26	0.01	7.72	5.53	4.64	4.14	3.82	3.59	3.42	3.29	3.18
27	0.05	4.21	3.35	2.96	2.73	2.57	2.46	2.37	2.31	2.25
27	0.025	5.63	4.24	3.65	3.31	3.08	2.92	2.80	2.71	2.63
27	0.01	7.68	5.49	4.60	4.11	3.78	3.56	3.39	3.26	3.15
28	0.05	4.20	3.34	2.95	2.71	2.56	2.45	2.36	2.29	2.24
28	0.025	5.61	4.22	3.63	3.29	3.06	2.90	2.78	2.69	2.61
28	0.01	7.64	5.45	4.57	4.07	3.75	3.53	3.36	3.23	3.12
29	0.05	4.18	3.33	2.93	2.70	2.55	2.43	2.35	2.28	2.22
29	0.025	5.59	4.20	3.61	3.27	3.04	2.88	2.76	2.67	2.59
29	0.01	7.60	5.42	4.54	4.04	3.73	3.50	3.33	3.20	3.09
30	0.05	4.17	3.32	2.92	2.69	2.53	2.42	2.33	2.27	2.21
30	0.025	5.57	4.18	3.59	3.25	3.03	2.87	2.75	2.65	2.57
30	0.01	7.56	5.39	4.51	4.02	3.70	3.47	3.30	3.17	3.07
40	0.05	4.08	3.23	2.84	2.61	2.45	2.34	2.25	2.18	2.12
40	0.025	5.42	4.05	3.46	3.13	2.90	2.74	2.62	2.53	2.45
40	0.01	7.31	5.18	4.31	3.83	3.51	3.29	3.12	2.99	2.89
50	0.05	4.03	3.18	2.79	2.56	2.40	2.29	2.20	2.13	2.07
50	0.025	5.34	3.97	3.39	3.05	2.83	2.67	2.55	2.46	2.38
50	0.01	7.17	5.06	4.20	3.72	3.41	3.19	3.02	2.89	2.78
60	0.05	4.00	3.15	2.76	2.53	2.37	2.25	2.17	2.10	2.04
60	0.025	5.29	3.93	3.34	3.01	2.79	2.63	2.51	2.41	2.33
60	0.01	7.08	4.98	4.13	3.65	3.34	3.12	2.95	2.82	2.72
100	0.05	3.94	3.09	2.70	2.46	2.31	2.19	2.10	2.03	1.97
100	0.025	5.18	3.83	3.25	2.92	2.70	2.54	2.42	2.32	2.24
100	0.01	6.90	4.82	3.98	3.51	3.21	2.99	2.82	2.69	2.59
200	0.05	3.89	3.04	2.65	2.42	2.26	2.14	2.06	1.98	1.93
200	0.025	5.10	3.76	3.18	2.85	2.63	2.47	2.35	2.26	2.18
200	0.01	6.76	4.71	3.88	3.41	3.11	2.89	2.73	2.60	2.50
1000	0.05	3.85	3.00	2.61	2.38	2.22	2.11	2.02	1.95	1.89
1000	0.025	5.04	3.70	3.13	2.80	2.58	2.42	2.30	2.20	2.13
1000	0.01	6.66	4.63	3.80	3.34	3.04	2.82	2.66	2.53	2.43

Source: From *Biometrika Tables for Statisticians,* Vol. 1, by permission of the Biometrika Trustees.

Table 9 continued

	Degrees of Freedom for Numerator									
10	12	15	20	25	30	40	50	60	120	1000
2.24	2.16	2.09	2.01	1.96	1.92	1.87	1.84	1.82	1.77	1.72
2.61	2.51	2.41	2.30	2.23	2.18	2.12	2.08	2.05	1.98	1.91
3.13	2.99	2.85	2.70	2.60	2.54	2.45	2.40	2.36	2.27	2.18
2.22	2.15	2.07	1.99	1.94	1.90	1.85	1.82	1.80	1.75	1.70
2.59	2.49	2.39	2.28	2.21	2.16	2.09	2.05	2.03	1.95	1.89
3.09	2.96	2.81	2.66	2.57	2.50	2.42	2.36	2.33	2.23	2.14
2.20	2.13	2.06	1.97	1.92	1.88	1.84	1.81	1.79	1.73	1.68
2.57	2.47	2.36	2.25	2.18	2.13	2.07	2.03	2.00	1.93	1.86
3.06	2.93	2.78	2.63	2.54	2.47	2.38	2.33	2.29	2.20	2.11
2.19	2.12	2.04	1.96	1.91	1.87	1.82	1.79	1.77	1.71	1.66
2.55	2.45	2.34	2.23	2.16	2.11	2.05	2.01	1.98	1.91	1.84
3.03	2.90	2.75	2.60	2.51	2.44	2.35	2.30	2.26	2.17	2.08
2.18	2.10	2.03	1.94	1.89	1.85	1.81	1.77	1.75	1.70	1.65
2.53	2.43	2.32	2.21	2.14	2.09	2.03	1.99	1.96	1.89	1.82
3.00	2.87	2.73	2.57	2.48	2.41	2.33	2.27	2.23	2.14	2.05
2.16	2.09	2.01	1.93	1.88	1.84	1.79	1.76	1.74	1.68	1.63
2.51	2.41	2.31	2.20	2.12	2.07	2.01	1.97	1.94	1.87	1.80
2.98	2.84	2.70	2.55	2.45	2.39	2.30	2.25	2.21	2.11	2.02
2.08	2.00	1.92	1.84	1.78	1.74	1.69	1.66	1.64	1.58	1.52
2.39	2.29	2.18	2.07	1.99	1.94	1.88	1.83	1.80	1.72	1.65
2.80	2.66	2.52	2.37	2.27	2.20	2.11	2.06	2.02	1.92	1.82
2.03	1.95	1.87	1.78	1.73	1.69	1.63	1.60	1.58	1.51	1.45
2.32	2.22	2.11	1.99	1.92	1.87	1.80	1.75	1.72	1.64	1.56
2.70	2.56	2.42	2.27	2.17	2.10	2.01	1.95	1.91	1.80	1.70
1.99	1.92	1.84	1.75	1.69	1.65	1.59	1.56	1.53	1.47	1.40
2.27	2.17	2.06	1.94	1.87	1.82	1.74	1.70	1.67	1.58	1.49
2.63	2.50	2.35	2.20	2.10	2.03	1.94	1.88	1.84	1.73	1.62
1.93	1.85	1.77	1.68	1.62	1.57	1.52	1.48	1.45	1.38	1.30
2.18	2.08	1.97	1.85	1.77	1.71	1.64	1.59	1.56	1.46	1.36
2.50	2.37	2.22	2.07	1.97	1.89	1.80	1.74	1.69	1.57	1.45
1.88	1.80	1.72	1.62	1.56	1.52	1.46	1.41	1.39	1.30	1.21
2.11	2.01	1.90	1.78	1.70	1.64	1.56	1.51	1.47	1.37	1.25
2.41	2.27	2.13	1.97	1.87	1.79	1.69	1.63	1.58	1.45	1.30
1.84	1.76	1.68	1.58	1.52	1.47	1.41	1.36	1.33	1.24	1.11
2.06	1.96	1.85	1.72	1.64	1.58	1.50	1.45	1.41	1.29	1.13
2.34	2.20	2.06	1.90	1.79	1.72	1.61	1.54	1.50	1.35	1.16

Table 10 Critical Values for Spearman Rank Correlation, r_s

For a right- (left-) tailed test, use the positive (negative) critical value found in the table under significance level for a one-tailed test. For a two-tailed test, use both the positive and negative of the critical value found in the table under significance level for a two-tailed test, n = number of pairs.

	Significance level for a one-tailed test at			
	0.05	0.025	0.005	0.001
	Significance level for a two-tailed test at			
n	0.10	0.05	0.01	0.002
5	0.900	1.000		
6	0.829	0.886	1.000	
7	0.715	0.786	0.929	1.000
8	0.620	0.715	0.881	0.953
9	0.600	0.700	0.834	0.917
10	0.564	0.649	0.794	0.879
11	0.537	0.619	0.764	0.855
12	0.504	0.588	0.735	0.826
13	0.484	0.561	0.704	0.797
14	0.464	0.539	0.680	0.772
15	0.447	0.522	0.658	0.750
16	0.430	0.503	0.636	0.730
17	0.415	0.488	0.618	0.711
18	0.402	0.474	0.600	0.693
19	0.392	0.460	0.585	0.676
20	0.381	0.447	0.570	0.661
21	0.371	0.437	0.556	0.647
22	0.361	0.426	0.544	0.633
23	0.353	0.417	0.532	0.620
24	0.345	0.407	0.521	0.608
25	0.337	0.399	0.511	0.597
26	0.331	0.391	0.501	0.587
27	0.325	0.383	0.493	0.577
28	0.319	0.376	0.484	0.567
29	0.312	0.369	0.475	0.558
30	0.307	0.363	0.467	0.549

Source: From G. J. Glasser and R. F. Winter, "Critical Values of the Coefficient of Rank Correlation for Testing the Hypothesis of Independence," *Biometrika, 48,* 444 (1961). Printed by permission of Biometrika Trustees.

Other Useful Tables

Table 8-2 Some Levels of Confidence and their corresponding Critical Values

Level of Confidence c	Critical Value z_c
0.75	1.15
0.80	2.38
0.85	1.44
0.90	1.645
0.95	1.96
0.98	2.33
0.99	2.58

Critical Values of z in Hypothesis Testing

Type of Test	$\alpha = 0.05$	$\alpha = 0.01$
Left-tailed Test	$z_0 = -1.645$	$z_0 = -2.33$
Right-tailed Test	$z_0 = 1.645$	$z_0 = 2.33$
Two-tailed Test	$z_0 = \pm 1.96$	$z_0 = \pm 2.58$